The Sun, The Cycles, The Science, and the Seers

Michael Lupo

Dedication

This work is dedicated to God, creator of the Universe. To my wife, Maddie, for allowing me the latitude to do the research in spite of her life challenges. I love you. This work is also a testament to two especially important research authors who have contributed an enormous amount to the ideas that this book is built upon. The two gentlemen whom I wish to give a special acknowledgment to have made this book go from a dream to a reality. The world has experienced a great loss in their passing over the last few years, and this book is a work of love done in honor of the late greats:

Erik Von Daniken

Courtesy:

https://www.amazon.es/Chariots-Gods-Erich-von-Daniken/dp/0285629115

Douglas B. Vogt

Courtesy:

https://www.youtube.com/watch?v=NFBGxi_InoI&list=PL1yQT0VIlSJkDD9n7Vr1MzIZH3upgSl6d&index=5

May we always enjoy the work that you both have left behind, find ways to advance the sciences because of your contributions, and we shall see you on the other side, my dear scientific colleagues.

Foreword

The rigors of greater knowledge require an open mind and flexibility of one's frame of reference so that all possibilities can be considered. Without this openness to learning innovative ideas, the insights that this book is about to share will be overwhelming and extremely hard to grasp. What the reader will encounter is an attempt to bring together a unified process. This unification is a consideration of the works of the authors who will be listed throughout the various parts of this book. When these writers are considered in totality, there is a purpose, flow, and a beauty of thought that helps one to see creation and the creator in a more meaningful way with greater understanding that leads to a deeper relationship with them. Some might call these writers a rabbit hole of conspiracy theories and pose accusations of pseudoscience. Others will vehemently argue the demonic. While there is a place for conspiracy theories and evil, this book is not that place. Let us take a moment to remember just how far Church thought has moved over time by considering the works of Copernicus.

The Catholic Church did not punish Copernicus directly for his heliocentric theory. Condemnation of heliocentrism came after Copernicus in 1616, during the Roman Inquisition, which was prompted by Galileo's zealous advocacy. The Church officially declared the heliocentric theory to be *"philosophically false and theologically heretical"*. All those things happened to the works of Copernicus because Galileo Galilei began publicly defending and promoting the ideas of Copernicus, which stated that heliocentrism was a physical fact, not just a hypothesis. This idea was seen by some in the Church as a direct contradiction of Scripture during the height of the Protestant Reformation. Because of the condemnation, the Church placed Copernicus's book *"De revolutionibus"* on an Index of Prohibited Books. This index is a notable example of narrative control that is exercised as an effort to control others. The book remained banned until it was published with corrections in 1620. Finally, in 1633, the Inquisition prosecuted Galileo for his continued support of Copernicus's theory, a much more famous and direct confrontation between science and religion that retroactively swept up Copernicus's work in the controversy, and the works of Copernicus were not removed from this Index entirely until 1835.

Today, the foundations of how we understand most of the solar systems that we observe in the night sky are based on the ideas of heliocentrism, where the various planets revolve around a central star, as initially discussed in Copernicus's work. Let us begin with the sun, how it works, and what is really happening to our solar system.

Table of Contents

PART I
The Sun

Introduction

Let us begin with these seemingly impossible ideas of a solar nova/pole shift, how they happen, and some of the mechanisms involved in the triggering of these events. There are forces exerted upon all celestial objects from several sources that do cause a repeating cycle of Solar Nova, and subsequent magnetic pole shifts. When our sun goes nova, the subsequent planetary reaction will be a pole shift on Earth. The main reason that mainstream science does not accept this premise is that physical evidence is so devastated all over the Earth from previous solar events that it is practically non-existent. Along with ancient evidence, there are near-extinction events of humanity that reset civilizations many times over.

It is the repeated reset of humanity that has given us collective amnesia concerning these existential events. As such, the mainstream of scientific thought finds it hard to admit many things that ancient artifacts, writings, myths, and religions, when considered all together, tell us a vastly different story concerning our past. As I am not an expert in any of these fields, I am simply a messenger bringing disparate thoughts and theories together so that a bigger picture might emerge concerning how our very distant past is showing us what may be to come. With that in mind, the following is a list of works that must serve as a starting point, which in no way is an exhaustive or complete list of researchers and theorists. There are other writers who have equally contributed to the idea of what exactly the sun does; this list is simply in alphabetical order by last name as follows:

1. Chris Bledsoe
2. Randall Carlson
3. Edgar Cayce
4. Ben Davidson
5. August Dunning

6. Graham Hancock
7. J. Allen Hynek
8. Matthew LaCroix
9. William J. Meister
10. Juan Moritz
11. Dianne Pasulka
12. Douglas Petrovich
13. Robert Schoch
14. Robert K.G. Temple
15. Jacques Vallée
16. Erich Von Däniken
17. Doug Vogt

We are going to begin with the combined ideas of nearly everyone mentioned above. Remember that most of these theories are extremely theoretical. What this means is that we really have no idea what will happen. We can infer the outcomes based upon records, evidence, and geologic history as to what is likely to occur during and after the major events of a micronova and pole shift for our great lake, planet Earth. The challenge for the next generation is to prove these ideas to be true. Our problem as the global family of homo sapiens in modern times is that we do not know what we do not know. Our history does not reach far enough with any clarity to say for sure what we as a people have experienced in our very distant past. For these and other reasons, we as a people are extremely critical of ideas that call into question what is considered outside of mainstream scientific understanding. This is also why some religious beliefs are bound in tradition, dogma, and cultural ideologies that do not allow for dissenting positions that challenge the status quo. There are several research and theoretical approaches that, when considered in tandem as distinct parts of the same process, come together to demonstrate a bigger picture of a very violent past, and of a rather promising future. It is precisely that violent past that is on the horizon to repeat concerning the sun, our galaxy, and how the interplay between them brings about existential threats that we can prepare for, live through, and survive.

By taking all the information contained in this book as a comprehensive approach so that some of us may likely survive the coming end of the age, this information is meant to empower people to make the appropriate spiritual, physical, and emotional preparations. First up is a classic theorist, the late great Douglas B. Vogt.

Chapter 1
Doug Vogt and The Sun

Throughout history, scientists and researchers have sought to understand the forces that shape our solar system and influence events on Earth. While many explanations rely on established scientific models, some theories explore more extreme possibilities involving the Sun itself. One such perspective is presented by Douglas B. Vogt, whose work examines the idea of recurring solar events and their potential impact on Earth. The following section introduces his theory and its key concepts.

Vogt suggests that the Sun undergoes a process where it superheats and sheds a dust shell, causing it to go nova. This event, according to Vogt, triggers significant and destructive changes on Earth. The following is a brief of his theory:

Solar Nova: The core of Vogt's theory is that the Sun experiences a recurring "nova-like" event where it blasts off a dust shell at the end of each long solar cycle, which repeats approximately every 12,068 years.

Pole Shift: This solar event, as described by Vogt, is the driving force behind Earth's polar reversals and the cause of ice ages.

Earth's Rotation Changes: The theory also proposes that these solar events affect Earth's rotation, leading to catastrophic changes in the planet's atmospheric patterns and potentially other geophysical shifts.

Vogt links these solar phenomena with religious concepts, interpreting these cyclical events in his book, "God's Day of Judgment". His book outlines a claim to have found the exact date for the next polar reversal, between September and December 2046, encoded in the Torah. He also claims the Torah confirms the 12,068-

year cycle. Vogt goes on to suggest that another of his books, *"Theory of Multidimensional Reality, an information theory of existence,"* explains the mysteries of the ice age. Vogt's books offer insights into the purpose of the Universe and man's place within it. The polar reversal is caused by a clock cycle that runs through time according to his theories on the sun. This clock cycle crosses the x-axis every 12,068 years. Some might call this a zero-point energy event. A complete cycle would represent two polar reversals, or 24,136 years. The magnetic field of the Earth will start decaying within 50 years of the actual polar reversal, but the decay will begin to exponentially increase as the Earth reaches the point of no return at seven years prior to the reversal of the poles, according to Vogt's writings.

The magnetic field does not have to go to zero before it snaps to a reversed polarity. As the magnetic field decays, it will create increased heat in the core of the Earth. The additional heat will rise towards the surface and manifest itself in the form of increased volcanoes and earthquakes. The increase in earthquakes is a result of the continental plates sliding more easily against each other. The increased heat of the mantle of the Earth, upon which the plates *"float,"* will become more lubricated, enabling them to move or fracture more easily. The increase in volcanic activity goes along with that process. The Earth's rotation will start slowing down in this period of time, resulting in the necessity to add seconds, then minutes, to our clocks according to Vogt's theory.

Vogt believes that weeks, and maybe months, before the reversal, the Earth's rotation will slow down markedly, maybe resulting in 28-hour days. The Sun will also be affected by its own collapsing magnetic field. The Sun's output will start increasing over 140 years before the next reversal. The increased solar output will start heating up the surface of the Earth. Sea surface temperatures will rise during sunspot cycles, successively increasing as we get closer to the reversal. The polar ice caps and glaciers will start melting, resulting in ocean levels rising. Vogt suggests that before the polar reversal, most of the ice caps and glaciers may melt. The increased sea surface temperature will create more violent and frequent storms worldwide. The increase in solar output will create additional levels of ultraviolet light hitting the Earth. Vogt's theory suggests that the ozone layer will deplete in the upper atmosphere during the sunspot maximum.

The Polar Reversal: The actual polar reversal occurs in one day. There is a complex series of events that will happen on that day. It all begins with what happens to the Sun.

What Happens on and in the Sun: The collapsing magnetic field deep inside the Sun will create a large spiking of energy that will cause the matter and dust shell on the surface to expand very rapidly. This event we would call a stellar Nova. Vogt estimates the speed of the dust shell to approach 1,550 miles per second. His best estimate for the energy output during the Nova is no less than 2,000 times the normal energy output.

The equatorial region of the Sun will blow outward along the planetary plane, hitting each planet as the dust shell expands rapidly. As the hot dust and matter shell hit each successive planet, it will push the planets a little further away from the Sun. After the Nova, the planets will receive a little less energy from the Sun. The planets also lose some of their atmosphere and surface liquids. Eventually, the dust/matter shell will lose enough momentum and stall out somewhere past Neptune. This is called the Kuiper Belt. After the Sun's matter shell has been expelled, it will give off mostly ultraviolet light but not that much radiant heat. The Sun will remain that way until the matter shell can form around it again, and that may take a number of sunspot cycles.

The Rotation of the Earth: At the exact time of the polar reversal, the Earth will stop its rotation and remain still for seven to eight hours. The forests and buildings on the sun side of the Earth will catch on fire if the plants are not flooded by the ocean or a nearby lake. Not only will heat affect plants and animals, but the sun will produce a massive dose of cosmic and gamma rays, which will reach the Earth within 10 to 15 minutes after the reversal. This cosmic particle pulse may last 10 to 30 seconds with the potential to alter the DNA of both plants and animals, including humans. This gamma burst is what creates new species based on older species, according to Vogt.

The Solar Dust Shell: The people on the sun side of the Earth will be able to see the sun expanding and the solar disk getting larger as it expands towards the Earth. Vogt estimates that the dust shell will take between 17 and 18 hours to hit the Earth. The dust shell will not necessarily hit the side of the Earth that faced the Sun at the moment of the polar reversal. This is because 18 hours will have passed, and another side of the Earth will be facing the Sun. When the dust shell hits us, it will deposit vast quantities of dust and rock on one-half of our Earth.

Solar Evaporation of Ocean Water and the Snow: The Nova will also evaporate at least six hundred feet of ocean water worldwide. Some of this water will be lost into space, carried away by the expanding dust shell. Some of it will appear to us as returning comets. Most of the water will remain in the atmosphere as superheated water for several days to a week. As the atmosphere cools down, it will start to rain hot rain.

As it cools further, the rain gets colder until it turns to snow. Lots of snow, and it will keep snowing for the next 40 to 50 years until most of the moisture is wrung out of the atmosphere. If you could see the Earth from space just after the Nova, you would see it with a thick tail, pointing away from the Sun. It will be made up of dust and ice crystals. Eventually, the debris tail will disappear over several thousand years, as the Earth's gravity brings the material back into the upper atmosphere.

Fast Frozen Animals on the Back Side: Just after the solar dust shell passes the Earth, the dust shell side will have extremely low atmospheric pressure because the dust shell would have blown most of the atmosphere away on that side; we just do not know how much. The backside of the Earth will temporarily have normal atmospheric pressure, but that condition will not last long. Very shortly after the dust shell passes by the normal atmosphere on the backside, it will expand very rapidly to fill up the front side or dust shell side of the Earth. Several things will result from this process:

- The first will be extremely high-speed winds traveling around the Earth from all four corners of the globe to fill up the extremely low atmospheric pressure on the dust shell side.
- The second consequence of this process is revealed in Boyle's law: *"If a given weight of gas is considered and if its temperature is held constant, the pressure and volume of the gas will be inversely proportional."*
- The third consequence is in applying this law to the conditions that will be present on the backside of the Earth.
- The atmosphere will expand extremely fast, and, therefore, the temperature will drop to extreme levels, possibly below -180 degrees Fahrenheit.
- Any life forms caught out in the open, or even in the average building, will be fast frozen almost instantly.

The Oceans: When the Earth stops its rotation, all the water in the oceans and lakes will continue to travel in the former easterly direction of the Earth's former rotational velocity.

Just before the reversal, the Earth may only be rotating at 800 mph, instead of the normal 1,000 mph at the equator, because the rotational rate will have slowed down considerably. Vogt compares it to walking at 5 miles per hour with a pan of water, and you stop abruptly. The water keeps going at 5 mph. He estimates that the wave speed will slow down considerably after the first hour on land. Many factors will determine the speed of the wave, such as land elevation, forest ground cover, gravity, and the heat blast from the Sun. The length of time the Earth remains at rest will determine how far this immense ocean wave will travel across the continents. When the Earth resumes rotation, it will be in the reverse direction from previously. This action will make the tidal wave appear to speed up once again, but really, it is because the Earth would now be rotating in the opposite direction to the wave. Vogt has estimated that the speed of this massive tidal wave would average about 350 miles per hour. Within a week, this massive ocean wave will return to the ocean basins. As it returns, it will naturally follow the river basins leading to the sea, cutting deeper into existing deep-sea canyons. Some deep-sea canyons are currently cut down to over 11,000 feet below sea level. The aftereffects are what some call slosh back, which is a series of smaller waves as the oceans settle back into their basins.

The Rotation of the Earth and Major Earthquakes: The rotation of the Earth should resume within seven or eight hours after the reversal point, but in the opposite direction. Currently, the Earth rotates from west to east.

After the reversal, it will rotate from east to west, so the Sun will rise in the west and set in the east. Not only are the oceans affected by the polar reversal, but also the crust of the Earth. Just as the oceans *"swim"* on the crust of the Earth, the crust of the Earth *"swims"* on the heated-up mantle. The continental plates are also traveling at the Earth's previous rotational velocity at the time of the reversal. They will start banging up against each other, thereby causing massive earthquakes lasting for weeks of constant shaking at over eight on the Richter scale until the plates stabilize. Mountains will be pushed up, and some may sink.

Volcanoes: Volcanoes will erupt worldwide because of the increased heat in the Earth's core and mantle and the pressure put upon the magma chambers in the crust. The dust from many erupting volcanoes will further add particles to the atmosphere. The volcanic dust will help seed the dense clouds already present because of the massive evaporation from the Nova. The lava and mud flows from these eruptions will also further reshape the surface of the Earth.

The Ice Age to Come: Within a few days of the Nova, it will start raining scalding hot rain, which will turn to downpours of cooler and finally chilly rain. Within 10 days, it should start snowing worldwide because the sun will not be giving off enough visible light and heat to warm the Earth. It will continue to snow until all the clouds and moisture are out of the atmosphere. This process may take 22 to 44 years or more. When it is done snowing, the Earth will be mostly covered with snow and thick ice in the higher latitudes and/or elevations. We will then be in the grips of a full ice age, maybe for a full seven hundred years, according to Vogt.

Life After These Events: The effects of the polar reversal on all forms of life will be devastating. It is not difficult to understand why there is always mass extinction of all types of species at the time of the polar reversal.

Vogt claims to have heard a program that suggested that biologists had figured out that there were only thirty-five females worldwide who survived the last ice age. He suggests that it is our job to survive the next ice age after these events occur in the 2040s. Vogt is imploring that, *"If you do not care about yourself, then think about your children or grandchildren. They should have a chance to survive this event."*

The Aftermath: Not all the Earth will get enough snow to form a glacier. Some areas near the equator will not be in the grips of a full ice age. It should be colder than the last ice age because we will be a little farther away from the sun. Dense clouds will cover most of the Earth for 22 or more years until all the moisture has been wrung out of it. The glaciers will block rivers and streams and create ice dams that will be present for hundreds of years.

When the ice dams finally collapse, they will cause huge floods on the rivers affected. Eleven to twenty-two years after the polar reversal, most of the snow and ice by the equators should have melted at the lower latitudes of 10-15° North and South. Animal life will start to come back in these areas. Small plants and trees will start to appear again. The oceans of the Earth will be over five hundred feet below the current sea level. Glaciers will be present in both the Atlantic and Pacific oceans and extend down to the lower North and South Latitudes. The occurrence of frequent major earthquakes should continue for at least 15 years after the reversal, until the outer mantle of the Earth cools down and the crustal plates stabilize.

Megatsunamis:

Doug Vogt believes that the Earth's rotation will go the other way as part of the pole shift from the nova event. It is this reversal that will cause the megatsunamis around the world. He also believes that if the dust shell impact evaporates more of the ocean, then the impacts will be less on the Pacific Ocean coasts; however, they will not be entirely mitigated.

Increased Run-up and Inundation: With less water buffering the wave's energy, a megatsunami would be able to travel further inland and reach higher elevations than under the present sea levels.

Intensified Erosion: The forceful flow of a megatsunami's leading edge, followed by the backwash, would lead to severe erosion of the exposed land, shaping the landscape in a manner that can remain evident for centuries.

Mountain Safety: Seek out areas at elevations of five thousand feet or higher, further inland than standard tsunami preparedness guidelines suggest, and near the equator. Pay close attention to natural warning signs and evacuate immediately to maximize the distance and height gained.

Intensified Erosion and Massive Sediment Transport: Be aware that the ground near the exposed continental shelf will be unstable and vulnerable to erosion, liquefaction, and landslides. Avoid areas where debris is likely to accumulate or where the ground may be unstable. Do not attempt to walk or drive through floodwaters, as debris can be deadly.

Potential for Enhanced Secondary Hazards: After a megatsunami, be prepared for potential secondary hazards such as fires, contamination of water sources, and hazardous material releases. Avoid wading in floodwater, as it can contain debris, sewage, and chemicals. Follow the guidance of local officials regarding the safety of water supplies and other potential hazards. Exposed, unstable sedimentary areas could be more vulnerable to liquefaction or landslides, which could trigger further tsunami waves or exacerbate existing damage.

Long-Term Geological Record: The unusual sedimentary patterns left by such a powerful event during lower sea levels could be preserved in the geological record for thousands of years, providing evidence of past tsunamis and allowing scientists to study their impacts in detail.

Significant Ecological Disruption: A megatsunami's impact would also extend to the environment, destroying habitats, killing animal and plant life, and potentially contaminating freshwater resources.

Massive Sediment Transport and Deposition: Large volumes of sediment, including sand and gravel, would be eroded from both the offshore and exposed continental shelf, and carried inland, eventually being deposited in thick layers over the existing landscape.

Scouring and Channel Formation: The backwash phase of a megatsunami, where the inundated water returns to the sea, can be particularly destructive, creating deep channels and scours as it flows down slope. These events will move an exceptionally large amount of sediments back towards the sea and will likely create a large inland sea in the central USA where the Mississippi River runs today.

Tsunami Safety Tips:

Tsunami Watch: A tsunami is possible; stay aware and be prepared to act.

Tsunami Advisory: A tsunami with dangerous currents or waves is expected or occurring; stay out of the water and away from beaches and waterways.

Tsunami Warning: A dangerous tsunami is expected or occurring; take immediate action and move to high ground or inland. Never go to the beach to watch a tsunami; if you can see it, you are too close. Stay away from the coast until officials announce that it is safe to return. A megatsunami during a period of lower sea levels that occurs because of a shift in the magnetic poles of the Earth after a nova event would be a geological and ecological force of immense destructive power, profoundly reshaping the exposed coastal and shelf environment and leaving a legacy in the geological record. The safest places are higher elevations that are not volcanic in nature, and that will be within 10 degrees of the equator after things settle down. This is because one of two things happens at and near the equator:

None to little ice forms as the rain cannot freeze, and so it never snows at or near the equator.

The accumulated ice sheet from snowfall will be much thinner at and near the equator, and so will melt in much less time than the rest of the planet.

Doug Vogt proposes that one organizes food for storage for around 30 or 40 years until you can get out of your cave or find ground that is warm enough with sunshine to grow plants again. This is when the galactic and solar cycles start over again. Vogt suggests that we save plants, animals, and some pollinating insects on a national scale. Vought explains that you must save those whom you can save and will be healthy enough to live through the solar events and ice age to come.

Chapter 2
Ben Davidson and The Sun

In the effort to understand the forces that influence the Sun and its behavior, some researchers look beyond internal processes and consider the role of larger galactic interactions. These perspectives suggest that the Sun does not operate in isolation but is instead affected by external cosmic forces that may trigger significant solar events. Ben Davidson presents a model that explores how galactic-scale energy, matter, and magnetic interactions could disrupt the Sun's stability. The following section examines his interpretation of these processes and the potential consequences they may have for Earth and the solar system.

Immediate Effects on the Sun:

Increased Accretion: The Sun would rapidly pull in vast amounts of interstellar material, significantly boosting its mass.

Intense Solar Activity: The added mass and magnetic chaos would trigger immense energy releases, causing:

- Massive solar flares and coronal mass ejections (CMEs).
- Dramatic increases in X-ray and UV radiation.
- A supercharged, turbulent solar wind.

Magnetic Reorganization: The influx of charged plasma and dust would disrupt and then violently re-establish the Sun's magnetic field, creating intense reconnection events and shockwaves.

Effects on the Earth & the Solar System:

Extreme Radiation: Earth's magnetic field would be overwhelmed, leading to massive aurorae and dangerous cosmic ray exposure, far exceeding current levels. This is a major contributing factor to the pole shift, and the Earth will tilt 90 degrees as the poles go from excursion event to full pole shift. This will move the USA to the southern hemisphere, place Canada / Alaska at or near the equator, place Siberia at the new north pole, move Antarctica to the equator, and place South America at the new south pole.

Atmospheric & Climate Chaos: Increased solar radiation and particle bombardment could strip planetary atmospheres and drastically alter climates.

Visibility: The solar corona would brighten, and dust clouds might become visible, potentially obscuring stars.

Long-Term Changes: Over time, the Sun's increased mass could slightly alter planetary orbits, though the main impact would be the intense, short-term disruption. Another is that some of the atmosphere surrounding the planets will be blown off and lost forever.

In essence, it would be a dramatic, chaotic, and destructive event, essentially a massive, continuous *"starquake"* on our Sun. The actual heliospheric current sheet is a very thin, low-density surface within our solar system, generated by the Sun's rotating magnetic field, and it does not accumulate significant matter. As the current sheet of the solar system interacts with the larger galactic current sheet, it is possible for a large amount of interstellar matter to build up upon the Sun.

Mainstream Science:

Increased Accretion: The Sun's gravity would pull in the material, and as this material, like dust, gas, and plasma, fell to the solar surface, the energy of its fall would be converted into heat and radiation.

Increased Brightness and Radiation: The extra infalling material would cause a significant, possibly rapid, increase in the Sun's luminosity and radiation output.

Magnetic Field Interactions: The influx of massive amounts of plasma would interact dynamically with the Sun's complex magnetic fields, potentially causing an extreme number of powerful solar flares and coronal mass ejections (CMEs), far exceeding those seen during a normal solar maximum.

Increased Density and Pressure: Rapid accumulation of matter onto the Sun would dramatically increase the density and pressure in its outer layers, potentially interfering with the standard processes of the solar dynamo and energy transport. This is understood to be a major contributing factor to the solar nova.

Magnetic Field Disruption: The influx of highly charged plasma could disrupt the Sun's magnetic field lines, leading to increased complexity and instability. The interaction of external magnetic fields with the Sun's own magnetic field could lead to massive solar flares and coronal mass ejections (CMEs) of unprecedented scale.

Solar Outburst/Micronova: In the context of the Davidson model, the extreme energy release from this rapid accumulation and magnetic reconnection would result in a massive, short-duration stellar explosion on the Sun's surface. This event would release an immense amount of energy and matter into the Solar System.

Chapter 3
Exothermic Core-Mantle Decoupling (ECMD) & The Dzhanibekov Oscillation (ECDO)

In physics, certain observed behaviors challenge our intuitive understanding of how objects move and interact, especially in environments like space. Some phenomena reveal that under specific conditions, rotating bodies can become unstable and shift in unexpected ways. One such effect is the Dzhanibekov effect, which demonstrates how rotational instability can lead to sudden changes in orientation. Building on this concept, some researchers have extended these principles to propose large-scale geophysical theories concerning Earth's internal structure and rotational behavior. The following section explores these ideas and the arguments surrounding them.

Imagine tossing a T-shaped handle in the air: while it spins, it can suddenly flip 180 degrees, even though you did not make it flip. This happens because the way the object's weight is distributed makes it unstable when spinning around a certain axis. In everyday terms, it is like watching a tennis racket or a phone tossed in the air suddenly twist or flip mid-spin for no obvious reason. There is a *"natural turnover"* of objects in the vacuum of space known as the Dzhanibekov effect. This is sometimes called the Tennis Racket Theorem or the Intermediate Axis Theorem. This phenomenon occurs when a rigid object with three distinct moments of inertia is spun around its intermediate axis, which is the axis that is neither the easiest nor the hardest to spin around. While spinning, the object will periodically and suddenly flip 180 degrees before returning to its original orientation. The Dzhanibekov effect happens because of rotational instability. According to Euler's equations for rigid body motion, rotation around the major or shortest, and minor or longest axes is stable; however, rotation around the middle axis is inherently unstable. Euler's equations are a set of mathematical formulas that describe the rotational motion of a rigid body around its center of mass. These equations account for how the angular velocity of an object changes over time when subjected to external torques. These equations take into consideration the object's moments of inertia along its principal axes.

They are fundamental in physics and engineering for analyzing the dynamics of spinning objects, such as planets, gyroscopes, or tumbling spacecraft. The primary reason for this effect is micro-disturbances. In the frictionless and weightless environment of space, even the tiniest wobbles amplify quickly, causing an object to *"tumble"* or flip end-over-end to conserve its angular momentum and kinetic energy. This was famously observed by Soviet cosmonaut Vladimir Dzhanibekov in 1985 when a spinning wing nut periodically flipped its axis while floating in the Mir space station. This effect gets its name from Dzhanibekov, as he was the first known demonstration of this oscillatory direction change. There are more common examples, like a tennis racket. If you toss a racket into the air while spinning it around the axis that runs through the face, it will almost always perform a half-twist before you catch it. Another group is T-Handles and wing nuts.

These are often used in NASA microgravity demonstrations to show how an object flips without any external forces being applied. When you scale this effect out to the planetary level, you get what some researchers call *"The theory of Exothermic Core-Mantle Decoupling"* (ECMD), and *"The Dzhanibekov Oscillation"* (ECDO). These theories propose a catastrophic, periodic geophysical event where Earth's inner layers decouple from the crust, causing rapid, large-scale shifts in the planet's rotation, axis, and magnetic field. These theories are based on interpretations of the *"Dzhanibekov Effect"* or intermediate axis theorem, combined with ancient evidence. These theories posit several effects as follows:

The "Flip" or Dzhanibekov Effect: Is the planetary scale effect on the Earth, which has an intermediate principal moment of inertia, it could behave like a tumbling T-handle in space, leading to a 180-degree flip of the entire crust/mantle, or a *"true polar wander"* event.

Stability of the Earth: Most physicists argue that the Dzhanibekov Effect does not apply to Earth because the planet is nearly spherical and rotates around its axis of maximum moment of inertia, which is inherently stable. Additionally, Earth's liquid interior and internal energy dissipation would likely suppress such a flip.

It is established that the Earth's inner core rotates at a different speed than the surface and that this rotation can reverse over several decades, which is a 60–70-year cycle. Mainstream physics and geophysics generally view the *"Earth flipping"* claim as unsupported or physically improbable because of what is commonly called the stability of the rotation of the Earth. The Earth is an oblate spheroid rotating on its maximum moment of inertia axis, making it stable against the Dzhanibekov or intermediate axis tumbling effect, which typically applies to asymmetric objects without stabilizing forces. There is an internal dissipation of energy or precession relaxation that works against the Dzhanibekov effect, keeping the Earth rotating stably around its axis of maximum moment of inertia. The geophysical hypothesis uses the ECDO to connect classical mechanics with ancient flood narratives to show past events and to predict future cataclysms.

Core Mechanisms:

Exothermic Phase Changes: The theory suggests that chemical reactions in the Earth's core release sudden bursts of heat or exothermic energy, which destabilize the thermal and gravitational link between the core and the mantle.

Decoupling: This heat is said to reduce mantle viscosity or melt the interface, causing the *"coupling"* that keeps the core, mantle, and crust rotating as a single unit to fail.

Core-Mantle Decoupling: Exothermic phase changes in the nickel-iron core or changes in angular momentum cause the liquid outer core, and or solid inner core to spin independently or oppositely to the mantle and crust.

Geophysical Consequences: This decoupling is hypothesized to cause rapid, large-scale tectonic movements, sudden changes in the magnetic poles, sometimes called geomagnetic collapse, and global floods. Proponents often point to the *"rapid"* reversal of magnetic poles, ancient flood myths, and alleged alignment of ancient monuments to predict shifted axes.

The Dzhanibekov Oscillation: Once decoupled, the Earth's rotation is modeled after the Dzhanibekov Effect or *"Tennis Racket Theorem"*.

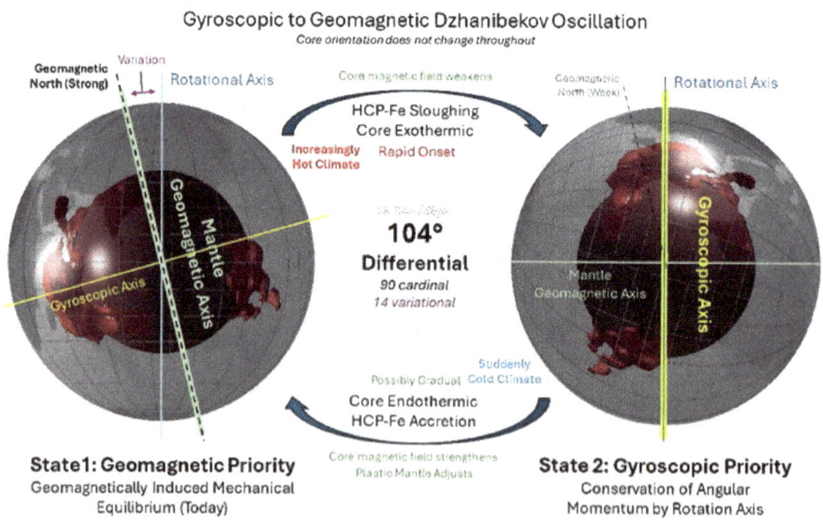

Courtesy: https://www.facebook.com/Piyacheep/posts/exothermic-core-mantle-decoupling-dzhanibekov-oscillation-theory-ecd-theory-is-/1283348816490894/

The Dzhanibekov effect is where a rotating object with three different moments of inertia becomes unstable and flips 180 degrees. Proponents argue the Earth's inner layers could "flip" or shift rapidly, leading to a Cataclysmic Pole Shift. Established science acknowledges that the Earth's inner core can rotate at a different rate than the surface, a phenomenon called differential rotation. Recent studies show the core's rotation may slow or even reverse slightly relative to the surface; thus, there continues to be accumulating evidence that the Dzhanibekov effect may also apply on a planetary scale.

Geophysical Evidence: Uses geological features and a sudden crustal slip to better explain tectonic movement.

Rapid Magnetic Excursions: Advocates suggest that mainstream geological features are better explained by sudden crustal slips than by gradual tectonic movement. Proponents point to Steens Mountain basalt flows, which contain magnetic signatures suggesting the magnetic field once changed by up to 3 degrees per day. While mainstream science often attributes this to anomalous cooling or localized magnetic flux, ECDO theorists use it to argue for rapid, large-scale shifts.

Anomalous Glaciation Patterns: A common argument that originated with Charles Hapgood is that certain regions, like North America, show evidence of deep glaciation that does not align with their current latitudes. They argue this is only possible if those landmasses were once located at the poles.

Sedimentary Megasequences: Large-scale, global sedimentary layers are cited as evidence of massive, inertia-driven floods that would occur if the Earth's crust suddenly shifted relative to the ocean's rotation.

Mantle Viscosity & Heat: Proponents utilize data on ultra-low velocity zones (ULVZs) at the core-mantle boundary.

ULVZs suggest that these dense, molten pockets are evidence of a *"lubricant"* layer that could facilitate crustal decoupling during exothermic events.

Historical Alignments of Ancient Monuments: A sizable portion of the theory relies on azimuthal misalignment of ancient structures, arguing that many sites were originally built to face a *"True North"* that has since shifted. Notable examples of azimuthal misalignment include the ancient temples of Teotihuacan in Mexico, where the main axis of the city and the Pyramid of the Sun are offset by about 15.5 degrees east of modern True North. Similarly, the Karnak Temple complex in Egypt is misaligned by approximately 27 degrees from the current north, while the megalithic site of Nabta Playa in southern Egypt is also oriented several degrees off True North. The Great Zimbabwe ruins in Africa and some of the megalithic temples in Malta exhibit recurring angular deviations, typically ranging from 5 to 30 degrees from modern True North. These misalignments support the argument that the builders originally oriented these structures to a True North that has since shifted due to a hypothesized pole shift event.

Systematic Misalignments: A sizable portion of the theory relies on azimuthal misalignment of ancient structures, arguing that many sites were originally built to face a *"True North"* that has since shifted. This historical alignment of ancient monuments is proof positive for some researchers who claim that hundreds of ancient sites are misaligned by specific, recurring angles that point to previous North Pole locations. One specific analysis states a claim that 38 monuments align to a predicted future or past pole with a statistically improbable p-value of $\sim 2\times10^{-140}$. A p-value of $\sim 2\times10^{-140}$ means the odds of something happening by random chance are astronomically tiny, and in fact so small that it is close to impossible. In everyday language, it is like saying if you rolled a die 140 times and got the same number every single time. The likelihood of that happening just by luck is practically zero. Statistically, it suggests that the result is not due to coincidence, but something real is going on.

The Giza Alignment: While the Great Pyramid is nearly perfectly aligned to the current True North, which is within 3/60ths of a degree.

Proponents argue that this is because the pyramids were built after the most recent shift. Older or *"pre-cataclysm"* sites are said to be the ones showing significant deviation.

Astronomical "Hinges": Sites like Stonehenge, Newgrange, and Chichen Itza encode precise solar and equinox alignments. Theorists suggest these sites were built by civilizations that understood the periodic nature of the Dzhanibekov Oscillation and designed monuments to track the shift of the *"cosmic hinges"*. Mainstream geophysics largely rejects these interpretations, noting that plate tectonics and isostasy or crustal buoyancy provide more consistent explanations for the geological record. Furthermore, fossil and paleomagnetic data do not show the global synchronized evidence required for a sudden 180-degree flip. What mainstream geophysics has failed to consider is that the flip is more like 104 degrees, and when triggered by a solar nova event, enough energy is released by the sun to exact such a sudden flip that otherwise would likely never occur on its own.

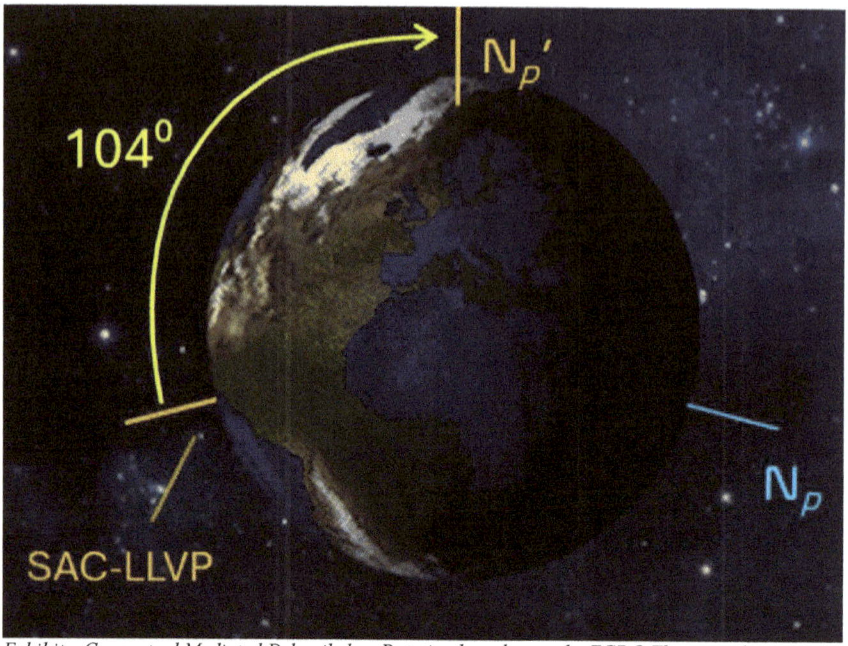

Exhibit – Conceptual Mediated Dzhanibekov Rotation based upon the ECDO Theory and established LLVP lower density axes within the Earth's mantle. This is known as a True Polar Wander (TPW) rotation and not a 'tilt' (change in obliquity). Courtesy https://theethicalskeptic.com/2024/05/12/exothermic-core-mantle-decoupling-dzhanibekov-oscillation-ecdo-hypothesis/

Current Geomagnetic Migration: The National Oceanic and Atmospheric Administration (NOAA) and the British Geological Survey recently released WMM2025, which confirms the following:

Unprecedented Drift: Since 1990, the magnetic north pole accelerated from 15 km/year to over 55 km/year.

Recent Deceleration: As of early 2026, the migration has slowed slightly to approximately 35–40 km per year.

Cause: This shift is driven by a *"tug-of-war"* between two massive patches of magnetic field, one under Canada and one under Siberia.

The ECDO Prediction: Proponents of the Dzhanibekov Oscillation model argue that the current magnetic drift is merely a *"precursor"* to a physical flip:

Rotational Instability: They posit that once the core and mantle *"decouple,"* the Earth will tumble like a T-handle in zero gravity until it reaches a new stable rotation axis.

Historical Precedent: Theorists often claim this occurs every 6,000 to 12,000 years, suggesting the *"next"* pole shift could end up in a location like Central Asia or the Southern Ocean, though specific coordinates vary wildly between authors.

Mainstream science tracks the Magnetic Pole, which is a moving point on a stable sphere. The ECDO theory predicts a shift of the Geographic Pole or the actual axis of rotation, an event for which there is no current empirical evidence in satellite or seismic monitoring. The argument that Earth's core has already begun to *"decouple"* rests on high-resolution seismic tomography that reveals anomalous behavior at the Core-Mantle Boundary (CMB). Proponents of the ECDO theory interpret these findings as signs of mechanical instability, while mainstream geophysicists view them as standard, and yes, complex fluid dynamics. The major issue is that the mainstream sciences are treating all this mounting evidence in the same way the Church treated Galileo concerning heliocentrism. Time is too short for arguments, and what we need is more open dialogue that moves empirical evidence towards greater understanding before the 2040s are here, so that at least some of us might have a chance to live through the coming end of the age.

The "Stalling" of the Inner Core: The most cited piece of data is a 2023 study published in Nature Geoscience, which analyzed seismic waves from repeating earthquakes.

Researchers found that the inner core's rotation, which previously bypassed the mantle's speed, appeared to *"pause"* or *"stall"* around 2009 and may now be rotating slightly in the opposite direction. ECDO theorists argue this *"pause"* is the first stage of decoupling, the moment the magnetic and gravitational *"clutch"* between the core and mantle slips, potentially leading to a Dzhanibekov-style tumble.

Ultra-Low Velocity Zones (ULVZs): Seismic monitoring has identified *"patches"* at the CMB where seismic waves slow down by as much as 30-50%.

These Ultra-Low Velocity Zones are located at the base of the mantle. Theorists suggest these zones are regions of partial melt or *"magma oceans"* that function as a lubricant. They argue that if these zones expand due to an exothermic heat burst, the friction holding the crust/mantle to the core would vanish, allowing the crust to slide, commonly called crustal displacement.

Large Low-Shear-Velocity Provinces (LLSVPs): There are two massive *"blobs"* under Africa and the Pacific, known as LLSVPs or *"Tuzo"* and *"Jason"*. These structures are thousands of kilometers wide and affect the Earth's gravitational symmetry. Because these blobs create an asymmetric mass distribution, theorists argue they function as the *"unbalanced weights"* necessary to trigger the Tennis Racket Theorem or Dzhanibekov Effect once the core-mantle coupling weakens.

"The Slow Sway": Mainstream science, such as recent research from the University of Southern California, suggests the core is not *"decoupling"* in a catastrophic sense. Instead, it oscillates back and forth over a roughly 70-year cycle. This oscillation is linked to minor changes in the length of a day by milliseconds rather than a 180-degree flip of the planet. Once again, the mainstream sciences have failed humanity because they are treating all this mounting evidence in the same way the Church treated Galileo concerning heliocentrism.

To model the *"lubrication"* of the Core-Mantle Boundary (CMB), one must calculate the energy required to transition the lower mantle from a solid or viscoelastic state to a low-viscosity liquid state across a global interface. Imagine the Earth as a giant jawbreaker. The core is the center, and the mantle is the outer candy layer. At the very bottom of the mantle, it is solid rock. To make the core *"decouple"* or slide smoothly from the mantle, scientists suggest you need a thin, liquid, lubricating layer that is like a layer of oil between two pieces of wood. To create this oil, you cannot just heat up the rock; you must melt it, which means turning solid mineral crystals like bridgmanite and ferropericlase into liquid. This process requires a specific, massive amount of energy called the *"latent heat of fusion"* just to break the solid structure apart, before the rock even gets hotter. The Core-Mantle Boundary (CMB) is massive, with a surface area that is extremely large. To calculate the surface area of the Core-Mantle Boundary (CMB) in square meters, you treat the Earth's core as a sphere and use the standard formula for the surface area of a sphere. The CMB is located at a depth of approximately 2,891 km below the Earth's surface. Given the Earth's mean radius is about, the radius of the outer core (the CMB radius). Then you plug the radius into the surface area formula: $A \approx 1.52 \times 10^{14}$ m^2. This is approximately 152,000,000,000,000 square meters. The spherical approximation is the standard method for general calculations.

To put this into perspective, the surface area of the Core-Mantle Boundary (CMB) is more than twice the total surface area of all the continents combined. For comparison, Earth's land area, including every continent and island, totals just over 148 million square kilometers, making the CMB one of the largest continuous interfaces inside the planet. This sheer size highlights the vast scale of geological processes occurring at this deep boundary.

Thickness: Even if you only melt a layer that is one kilometer or 0.6 miles thick, it will basically create a *"skin"* on the core compared to the thousands of kilometers of depth; therefore, you are melting a truly enormous volume of rock.

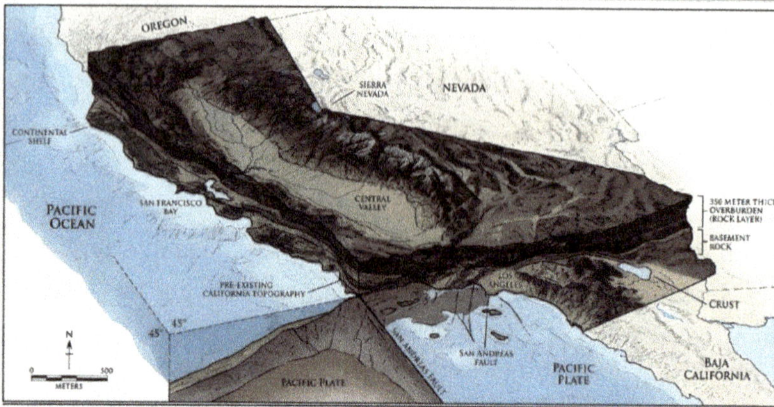

"GEOLOGICAL CROSS-SECTION: CALIFORNIA UNDER A 350M ROCK LAYER" (45° PERSPECTIVE)

The Result: You would need to melt approximately 150 trillion cubic meters or A≈1.52 x 1014 m2 of rock. To visualize in terms of continental size, imagine covering the entire state of California (about 424,000 km²) with a layer of rock roughly 350 meters (over 1,100 feet) thick.

Alternatively, if you spread this volume across the entire surface area of Texas or about 695,000 km², the layer would be around 215 meters that is about seven hundred feet) thick. This is an enormous volume, comparable to blanketing a large U.S. state with a skyscraper-tall sheet of solid rock.

The Heat: To turn that much solid rock into liquid at such high pressures requires a staggering amount of energy. Using the Latent Heat of Fusion for typical mantle rocks, you have $Q=mXL_f$.

In this formula, Q is the total heat energy required measured in Joules with a *J; m* is the mass of the rock measured in kilograms; and L_f is the Specific Latent Heat of Fusion for the rock that is measured in *J/kg or kJ/kg.*

The energy required to melt this 1-km layer is approximately the total daily energy consumption of humanity. By comparison, this is equivalent to roughly one billion times the energy released by the

Tsar Bomba, the largest nuclear weapon ever detonated. This is mathematically expressed as ≈1.8 exajoules.

The Exothermic Source: Proponents of the ECDO theory, such as those discussed in geophysical forums, suggest this energy comes from *"phase-change ripples"* or nuclear georeactors in the core. What both ECDO proponents and the mainstream sciences have failed to consider is the energy released from a solar nova as the triggering amount of required energy to make decoupling a reality on a global scale.

The Math Problem: For the core to release Joules suddenly, it would require a chemical or nuclear reaction of unprecedented scale. Mainstream thermodynamics, governed by the Second Law of Thermodynamics, suggests that heat transfer at the CMB is a slow, convective process, not an explosive one.

"Clutch" Mechanism or Viscosity Math: The mechanical coupling between the core and mantle is maintained by viscous drag and topographic coupling of mountains on the underside of the mantle sticking into the liquid outer core.

Decoupling Threshold: To trigger a *"slide,"* the viscosity of the boundary layer would need to drop from a solid-like to nearly liquid.

The Result: Mathematical models of Runaway Plate Tectonics suggest that if this lubrication occurred, the crust could theoretically move at meters per second rather than centimeters per year.

Centrifugal vs. Gravitational Forces: Once decoupled, the Dzhanibekov Effect math takes over. Because Earth is an oblate spheroid that is bulging at the equator, its rotation is highly stabilized.

To overcome the *"equatorial bulge"* and flip the axis, you would need to overcome the Earth's rotational kinetic energy. The Earth's rotational kinetic energy is an estimated calculated in Joules. This value is calculated using the formula for rotational kinetic energy, which is Earth's moment of inertia, or a formula to show its angular velocity, which is roughly demonstrated in radians per second. The math is more complicated than I can express in this book, and it is strongly recommended that you investigate the math involved to understand the various calculations for these effects. Basically, these are three orders of magnitude greater than the energy needed to simply melt the boundary. While the math for melting a lubricant layer is theoretically *"possible"* with enough hypothetical energy, the math for flipping the entire planet's axis requires energy levels typically reserved for massive planetary collisions.

Theorists often synthesize these two elements by linking real-time seismic anomalies to a cyclical *"doom clock"* derived from historical records. What mainstream science fails to connect is that a solar nova would provide that and more energy to make this a reality. And suddenly, Galileo seems to be the denial order of the day for many mainstream sciences that will, for their own pride, be the demise of us all if things do not change quickly, and we begin to seriously consider this as a real possibility.

Seismic Indicators of CMB Heating: Proponents argue that we are currently in the *"thermal buildup"* phase of the cycle, using deep-Earth data as evidence that the core-mantle boundary (CMB) is destabilizing.

Deep-Focus Earthquake Swarms: Theories like those found in the ECDO theory documentation suggest that an increase in deep-focus earthquakes that are below three hundred kilometers indicates rising thermal pressure from the core. They track data from the USGS Earthquake Catalog to argue that *"mantle plumes"* are becoming more aggressive.

The "Slowing" Core: As mentioned, the 2023 Nature Geoscience study on the inner core's rotational *"pause"* is used as the smoking gun. Proponents claim this is not a natural oscillation but the *"clutch slipping"* due to the core overheating and melting the boundary layer.

Expanding Geoid Anomalies: Data from the GRACE-FO satellite mission show shifts in Earth's gravity field. Theorists interpret these as mass redistributions at the CMB that will eventually function as the *"unbalanced weight"* triggering the Dzhanibekov flip.

Historical Timelines of Earth "Flip": ECDO theorists rely heavily on the 12,000-year cycle that is often divided into 6,000-year sub-cycles, aligning geophysical events with major historical disruptions. It is rather interesting that all the other researchers and scientists covered in this book see this very same cycle of galactic and solar time repeat, bringing an extinction-level event to the Earth on a consistent long timeline.

ECDO Timeline, Estimated Dates, Events, and Evidence:

Estimated Date	**Event in ECDO Timeline**	**Supporting Evidence Cited**
~10,500 BCE	The Great Flip	End of the Younger Dryas, mass extinctions of megafauna like mammoths, and the *"flood"* myths.
~4,500 BCE	The Intermediate Shift	Sudden desertification of the Sahara; the Ubaid period transitions in Mesopotamia.
Current Era	The 2025-2048 Window	Acceleration of the Magnetic North Pole and the *"stall"* of the inner core.

The "Clock" in Ancient Monuments: Theorists claim that ancient civilizations were *"post-cataclysm"* survivors who built *"clocks"* to warn us of the next decoupling:

The Zodiacal Ages: Proponents argue that the Precession of the Equinoxes is not just a slow wobble but a countdown. Each 2,160-year *"age"* represents a stage in the thermal loading of the core.

The Chan Hol Skeleton: Archaeological finds in flooded caves, such as those discussed by the National Institute of Anthropology and History (INAH), are sometimes used to suggest that sea levels rose so rapidly due to a *"crustal slip"* that entire populations were trapped instantly. In contrast, mainstream archaeologists and geologists note that these *"timelines"* cherry-pick data.

As an example, consider the carbon dating of the Sahara's green period. The results show a transition over centuries, not the days or weeks a Dzhanibekov flip would require. Once again, mainstream sciences have failed to account for pole excursions that are not full flips and can occur over hundreds to thousands of years between actual solar nova events. This is yet another example of the Galileo effect in the sciences.

Predicted "New Pole" Locations: Theorists typically predict the rotation axis will shift by roughly 40 to 104 degrees. In the ECDO and Dzhanibekov-flip communities, *"safe zones"* are calculated by projecting where the new Geographic North Pole will land and then identifying landmasses that avoid the resulting Inertial Flood Zones and New Arctic Circles.

"Safe Zones": According to fringe survivalist maps, a location is deemed "safe" based on three criteria:

Elevation (The 2,000-Foot Rule): Because a crustal slip would cause the oceans to *"slosh"* due to inertia, proponents argue you must be at least 2,000 feet or six hundred meters above sea level to survive the initial megatsunamis.

Tectonic Stability: Sites must be away from the Ring of Fire or major fault lines, as the decoupling is expected to trigger massive volcanic eruptions.

The "New Equator" Projection: Proponents look for regions that remain in the temperate zone after the flip. This is a key survival concept as the impending results are the next age of ice, and the temperate zones close to the new equator will be the first to melt, thus making them favorable for survivability to humanity.

High-Ground Refuges: The Altiplano, South America: Regions like Bolivia and Peru are favored because of their extreme altitude and distance from the predicted "slosh" paths of the Atlantic and Pacific.

The Tibetan Plateau is often cited as the ultimate *"Ark"* due to its average elevation of 14,000 feet. Historical references to Shambhala are sometimes used as *"encoded"* memories of previous survivors.

The Ozark Plateau/Appalachian Mountains: In North America, these are cited as secondary safe zones because they are geologically *"old and stable,"* though their safety depends entirely on whether the North American plate stays at a temperate latitude.

The Colorado Rocky Mountains: In North America, these are cited as a primary safe zone because they are geologically *"old and stable,"* though their safety depends entirely on whether the North American plate stays at a temperate latitude. The average height is well over two thousand feet.

The African Highveld: Much of South Africa and Ethiopia is considered safe by theorists because the African plate is *"anchored"* over a massive Mantle Blob (LLSVP), which some believe will function as a stabilizer during the flip.

The "Siberian" Pole Scenario: If the geographic pole moves toward the current magnetic drift (Siberia), the new Arctic Circle would encompass much of Russia and Alaska.

The "Bay of Bengal" Scenario: Some models, popularized by other researchers like Chan Thomas, suggest the pole could shift to the Indian Ocean. This would place the United States and Europe in a deep freeze, while Antarctica would become a tropical continent.

Likely Outcome: When looking at all of the theories and options discussed, it is most likely to be between 90 and 104 degrees of shift that puts most of the USA in the southern hemisphere with Alaska and Greenland at or near the equator, and Antarctica in the upper southern hemisphere also close to the equator. The new South Pole will be South America, and the new North Pole will be Siberia and China.

Mainstream Geophysics: The Geodetic Reference effect is supported by the NASA Global Differential GPS, which monitors the Earth's crustal movement in real-time. Current data shows the crust moving at roughly the speed that your fingernails grow, which is about 2–10 cm per year. There is no empirical evidence of the massive, kilometer-scale *"sloshing"* or rapid crustal displacement required to make these safe zones necessary, according to them. The preponderance of evidence collected here is screaming otherwise. If you do not consider alternatives, you will not live to see humanity through the coming end of the age. I have a sincere desire to give humanity the option to survive in the next age to come, and it is precisely these considerations that will give some of us the wherewithal to see humanity through it.

Chapter 4
Dr. August Dunning and the Sun

In the effort to better understand the behavior of the Sun and the events that may influence it, some researchers have explored the possibility that solar activity follows repeating patterns. These perspectives suggest that the Sun may experience periodic disruptions that have direct consequences for Earth and its environment. Dr. August Dunning presents a model that examines recurring solar nova events, their potential causes, and the implications they may have for planetary systems. The following section outlines his key ideas and observations.

Recurrent Nova Activity: According to Dunning, many stars, including our Sun, exhibit periodic nova events due primarily to thermonuclear runaways in their hydrogen-rich envelopes. These events can occur roughly every 12,000 years, affecting the Earth drastically.

Solar Outbursts: Dunning highlights that solar novae can lead to massive solar outbursts, which are brief, intense flashes from the Sun that can drastically alter Earth's climate and environment. These events can release enormous amounts of energy, acting as a sudden release mechanism that affects various layers of the Earth's atmosphere.

Geomagnetic Excursion: Dunning links these solar events to geomagnetic excursions or small shifts in Earth's magnetic field, suggesting that such excursions may be precursors to catastrophic events, which may lead to significant climatic and geological changes. One of which is a full magnetic pole shift.

Recent Observations: Dunning points to historical precedents, like the Younger Dryas event approximately 12,000 years ago, which was highly destructive and considered linked to past solar activities.

Modern Implications: He notes an alarming trend where the Earth's magnetosphere has been weakening, leading to increased vulnerability during solar activities. This has been observed as a 25%-30% decline in magnetic strength over just a couple of decades, potentially foreshadowing intense solar events. Dr. Dunning's assertions serve as a call to recognize the potential risks posed by solar novae events, urging awareness and preparedness in the face of these natural phenomena. Dunning's research integrates historical data with modern observations, indicating a need for continued study on solar activities and their implications for Earth.

Part I of this book outlines scientific theories from Doug Vogt, Ben Davidson, Dr. August Dunning, and others, all of which converge on the central theme of a recurring solar event that is most often referred to as a solar nova or micronova. The Solar Nova is a major contributing factor to catastrophic, cyclical changes on the Earth, which include geomagnetic excursions and pole shifts as well.

Key Connections:

Cyclicality of Events: All the covered scientific theories suggest that these catastrophic events occur in cycles. Vogt, Davidson, the ECDO, and Dunning specifically mention a cycle of approximately 12,068 years or 12,000 years, linking historical events like the Younger Dryas to past solar activity or global events that occur because of the solar nova.

Solar Activity as the Cause: The core idea is that changes within the Sun, like the collapsing of its magnetic field or the rapid accretion of interstellar material from the galactic current sheet, trigger massive energy releases, solar flares, CMEs, X-ray bursts, and yes, even solar nova/pole shifts.

Earth's Magnetic Field Weakening: A common precursor identified in these theories is the weakening of Earth's magnetic field. A weaker magnetic field on Earth makes the planet more vulnerable to incoming solar radiation and high-energy particles, which most likely cause climate change by heating both the inside and outside of the Earth.

Geomagnetic Excursions: The solar outbursts are linked to geomagnetic excursions and full pole shifts.

Pole Shifts: This is where the Earth's magnetic poles move or reverse, potentially causing the physical crust to shift and alter the planet's rotation and or continental locations in relation to the equator and poles.

Devastating Earth Impacts: The predicted consequences are largely the same for the various theories, like massive earthquakes, worldwide volcanic eruptions, rapid and extreme climate shifts from intense heat to a sudden ice age, massive ocean waves like mega tsunamis, and high-speed winds. These events lead to mass extinctions and the rapid *"flash-freezing"* of life forms on a global scale.

Radiation Exposure: All theories imply an increase in harmful radiation, like UV, cosmic, and gamma rays reaching Earth due to the compromised magnetic field, which may alter DNA and create new species.

Essentially, part one demonstrates the values of an integrated perspective into a single catastrophic model where a solar event triggers a chain reaction of geophysical changes, leading to a rapid and devastating reshaping of Earth's surface and biosphere. Doug Vogt, Ben Davidson, Dr. August Dunning, and the ECDO all converge on the idea of recurring, catastrophic solar events. These events are called a solar micronova, or solar nova, triggering rapid, devastating changes on the Earth, including either the geomagnetic pole or possibly physical pole shifts and mass extinctions.

Mainstream science largely views these specific, rapid-onset theories as unsupported by current empirical evidence; this, my dear reader, is a prime example of *"The Galileo Effect."* Mainstream sciences acknowledge the reality of magnetic pole migration and natural climate cycles that occur over much longer timescales, which support the theories in their totality that are gathered in the pages of this book. Do not take my word for truth; read the works of the various theorists listed in the intro and make your own conclusions.

Chapter 5
Lessons Learned So Far

When the ideas presented in the previous chapters are examined together, several consistent themes begin to emerge. Although each researcher approaches the subject from a unique perspective, their conclusions often point toward similar cycles, triggers, and outcomes. By bringing these viewpoints into a single frame, it becomes easier to identify the patterns that connect solar activity, planetary response, and historical events. The following section outlines the key lessons drawn from these combined perspectives.

Cyclical Catastrophes: All these theorists propose a catastrophic cycle of approximately 12,000 years, linking past events like the Younger Dryas to solar activity.

Solar Triggering Factor: The central concept is that changes within the Sun, e.g., a collapsing magnetic field or rapid accumulation of interstellar material, cause massive energy releases that bombard Earth with radiation and particles.

Earth's Magnetic Field: A key precursor identified is the weakening of Earth's magnetic field, making the planet more vulnerable to incoming solar radiation. Mainstream science confirms the magnetic pole is currently migrating rapidly, but does not link this to a sudden, catastrophic event.

Geomagnetic and Physical Pole Shifts: The solar outbursts are linked to geomagnetic excursions and full pole shifts, potentially causing the physical crust to shift. Mainstream science generally considers a rapid physical flip of Earth's rotational axis physically improbable due to the planet's stable rotation around its maximum moment of inertia.

Devastating Impacts: Predicted consequences across the theories include massive earthquakes, worldwide volcanic eruptions, rapid climate shifts from heat to a sudden ice age, megatsunamis, and high-speed winds, leading to mass extinctions and flash-freezing of life forms.

The ECDO and Dzhanibekov Effect:

The Exothermic Core-Mantle Decoupling Oscillation (ECDO) theory utilizes the *"Dzhanibekov Effect"* or tennis racket theorem to model a potentially catastrophic, periodic geophysical event where Earth's inner layers decouple from the crust.

Mechanism: Proponents argue that heat bursts in the core reduce mantle viscosity, allowing the core/mantle to decouple from the crust. Once decoupled, the planet could tumble up to 180 degrees to a new stable rotation axis.

Evidence: Theorists point to the current rapid magnetic pole drift, the 2023 study on the inner core's rotational "pause," anomalous glaciation patterns, and the azimuthal misalignment of ancient monuments as evidence for past shifts and a coming event.

Galileo Effect: When mainstream geophysicists and any number of other sciences largely reject the application of alternative ideas by stating things like, *"The Earth's near-spherical shape and internal energy dissipation make a rapid flip impossible,"* or *"The inner core rotation does oscillate but is not a sign of catastrophic decoupling."* These denials of the preponderance of evidence are: *"The Galileo Effect."* When a given mainstream science flat-out rejects an alternative theory or calls an alternative theory pseudoscience.

Survival Tips:

Part I outlines specific survival strategies for the predicted solar/pole shift events, which are primarily based on the following considerations:

- *Elevation:* Seek elevations of at least 2,000 feet or six hundred meters or higher to survive megatsunamis caused by the oceans' inertia during a rotational stop.
- *Location:* Move to stable, non-volcanic regions near the predicted new equator or at the lower latitudes within 10 degrees North/South, where ice melt would be faster and a full ice age might be avoided.
- *Tectonic Stability:* Avoid the Ring of Fire or major fault lines due to expected massive, weeks-long earthquakes and volcanic eruptions.
- *Specific Refuges:* Regions like the Altiplano of Bolivia/Peru, the Tibetan Plateau, the Ozarks/Appalachians, the Colorado Rockies, and the African Highveld are cited as potential *"safe zones"*.
- *General Preparedness:* Prepare for power outages, have an evacuation plan, and stay aware of natural warning signs like a loud ocean roar, receding water, and strong ground shaking.

These survival tips can be supplemented by government emergency services online at the following links:

- https://www.ready.gov/
- https://www.noaa.gov/

Both government sites recommend having an emergency kit, a family communication plan, and specific actions for known hazards like tsunamis, mega tsunamis, solar storms, solar nova, and pole shifts. It is rather interesting that these government entities have such suggestions, while at the same time, mainstream science is caught up in the Galileo Effect on many of these same ideas concerning the cycles of the sun and the aftermath that a solar nova will bring to the Earth. We are now going to move into Part II, where we will break down the science that the theorists have established behind the galactic, solar, and pole shift events to demonstrate that these ideas are much more than just similar; they are related to one another.

PART II
The Cycles & The Science

Introduction

In Part I concerning the Sun, you have been introduced to several writers, researchers, and scientific theories that express theoretical thoughts that science is beginning to bear out today concerning the cycles of the Sun. There are several associated cycles that play a major part in the solar nova/pole shift events. Part II will expand upon the underlying scientific evidence so that you might gain clarity with the why behind the reason that there should be more concern for long-term cyclical events. What you will gain is a deeper understanding of how those events have influenced life on Earth for much longer than the mainstream sciences care to admit. While there are some cycles that will not be touched upon in this book, they may be worth your reading consideration upon completion of this book. Part II begins with a high-level overview of the sciences involved and how some researchers and theorists have shown that, on a quantum level, things are more connected than we may have understood up to this point. We must take a deeper dive together into the underlying science associated with the solar events and cycles that have already been discussed. There are varying degrees of support, research, and evidence on galactic and solar cycles, and several scientists have demonstrated great insights that provide years of reading material for those interested enough to dig into their respective works. There are several authors, and in Part II, the list is essentially the same as those found in Part I.

While Doug Vogt got my wheels turning early on in my search for answers concerning the solar nova and pole shifts, I am now focused on three specific theories and the science behind the respective theories of each one concerning solar micro nova and pole shift events on Earth. The first is Ben Davidson, the second is August Dunning, and the third is the ECDO. While Mr. Vogt has sound ideas that we will cover in Part II, he did not have a defined approach to the pole shift, and what exactly may happen concerning the Earth's tilt and rotation. Both Davidson and Dunning consider the galactic current sheet and how our sun engages there over time. It is the electromagnetic current sheets along with the EDCO effect that seem to be worth the most consideration to date. The subsequent chapters of Part II will give the reader a better understanding of the foundations for these various scientific theories. The various experts listed below have an enormous amount of material that is a lifetime of recommended reading and videos to enjoy. Some of these authors are a repeat from Part I precisely because these theorists have contributed in important ways to both the first and the second parts of this book. The following list is in no way an exhaustive or complete list, as there are others who have equally contributed to the idea of galactic cycles, solar cycles, and extinction events that repeat over cosmic time. This list is simply in alpha order by last name as follows:

1. Chris Bledsoe
2. Randall Carlson
3. Edgar Cayce
4. Ben Davidson
5. August Dunning
6. Graham Hancock
7. J. Allen Hynek
8. Matthew LaCroix
9. William J. Meister
10. Juan Moritz
11. Douglas Petrovich
12. Robert Schoch
13. Robert K.G. Temple
14. Jacques Vallée
15. Erich Von Däniken
16. Doug Vogt

Chapter 6
History Reveals Cycles

We must begin with some lessons in galactic time periods that some sciences call cycles. *"How History Reveals The Cycles"* Is the work of Randall Carlson. His perspective on Earth's history emphasizes geological catastrophism and challenges the traditional gradualist narrative, proposing that recurring cataclysmic events, often cosmic in origin, have shaped the planet and human civilization.

Some of Carlson's Core Ideas:

Cyclical Catastrophism: Earth's history, and the rise and fall of civilizations, is shaped by predictable, recurring cycles of catastrophic change, not just slow, gradual processes.

The Younger Dryas Impact: A massive cosmic impact event that was most likely a cometary or asteroidal airburst around 11,600 to 13,000 years ago caused the Younger Dryas climate shift, a sudden return to ice age conditions that ended the previous interglacial warming.

Mega Floods and Geological Evidence: The Younger Dryas impact triggered immense mega floods like the Missoula Floods and megatsunamis.

These mega floods left behind significant geological evidence like massive boulder fields, altered landscapes such as Canyon Lake Gorge, and a carbon-rich *"black mat"* layer containing materials formed under extreme heat and pressure like nanodiamonds, microspherules, and iridium. We will learn more about this evidence when we discuss buckyballs, and the reason they are also found on the moon is because they are caused by a solar nova event.

Lost/Advanced Civilizations: The cataclysmic events wiped out or set back previous advanced civilizations, which is why the traditional historical narrative is incomplete. Remnants of their knowledge may be encoded in ancient myths, legends, and monumental structures around the world.

Ancient Knowledge and Cosmic Cycles: Ancient peoples were not primitive; they possessed sophisticated knowledge of astronomy, sacred geometry, and the cosmic cycles that govern these catastrophic events, a wisdom largely lost to modern culture.

Mathew Lacroix is another researcher of ancient civilizations, and according to him, there is evidence of these long cycles of events in Earth's ancient past. Lacroix offers a timeline to demonstrate this long-term cycle, and it makes for a great baseline from Earth to expand out from there to the solar system and then on to galactic time and cycles. It is not just coincidental that these timelines all align concerning the sun and subsequent ice ages over time, as a foundation for solar / pole shift events. While Mathew does not specifically call them out as ice ages, the event names are listed, like the Younger Dryas event. Lost civilizations are extraordinarily strong supporting evidence that something serious has happened and will occur again at the end of the age. This is not a matter of fear; however, it demonstrates an opportunity to choose love over fear as we learn the connections found between The Sun, The Cycles, The Science, and The Seers.

200,000 Year Timeline of Civilization

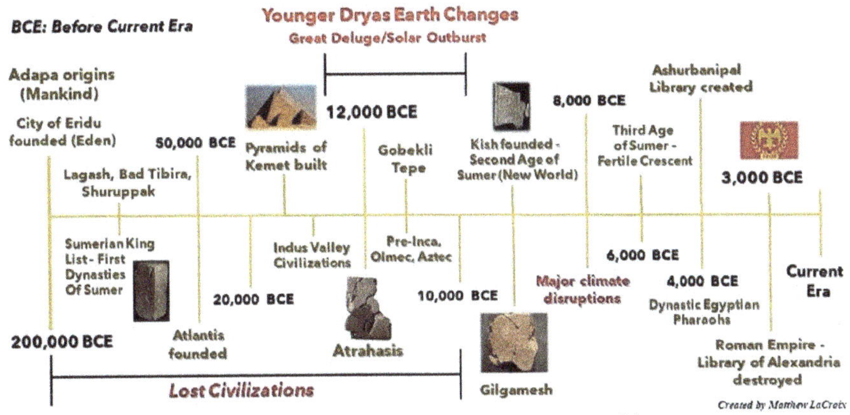

Courtesy of the works of Mathew Lacroix. Retrieved from https://www.youtube.com/watch?v=UU_L1YmrJjg

The next consideration is a deeper dive into a timeline that, according to Randall Carlson, shows how historical events are cyclical. I find it to be a synchronicity that Carlson's timeline outlines what he calls *"A Great Year,"* and Lacroix's timeline seems to fit together rather nicely. According to Carlson, a great year is a representation of the processional cycle of the Earth that divides the 25,920-year processionary cycle into four *"great seasons"* of 6,480 years each. Within his model, other key time units include a "Platonic month" of 2,160 years, a *"decan"* of 720 years, and a *"half great year"* of 12,960 years, which marks the reversal of seasons. This framework is based on the precession of the equinoxes, the slow wobble of Earth's axis.

Carlson does a fantastic job of establishing these movements and connecting them to cyclical events that, in most cases, are extinction-level events on a global scale here on Earth. This evidence shows that this universe is not a very easy place to live in, and so finding a way to reach the eternal becomes increasingly paramount with a greater understanding of natural processes that are so far beyond our control that they speak of a creator. We are a part of creation, and our small human minds cannot comprehend with ease or certainty, the grand scale of these timelines. The grandeur of these scales can be overwhelming, so do consider them prayerfully.

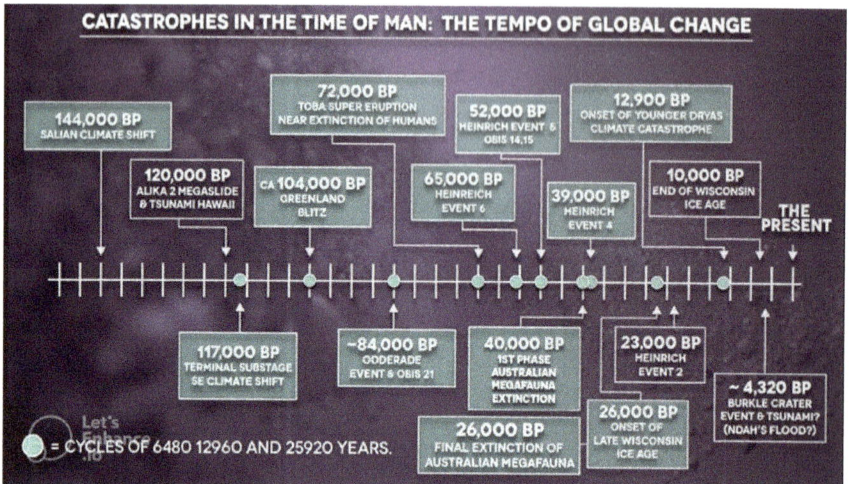

Courtesy of the works of Randal Carlson, enhanced by let's enhance. Io. Retrieved from, https://sacredgeometryinternational.com/the-great-year-of-the-world-sacred-geometry-cosmic-cycles-and-catastrophe/

If linear time is too confusing, Carlson also uses a circular option. The next graphic shows us this complexity in somewhat of a compass style or clock-like image to demonstrate the repeating nature of the cycles over time. The key here is an astronomical symbolism for time and not astrological, so all you fear mongers over divination can feel relieved and feel-good knowing that this graph is not evil.

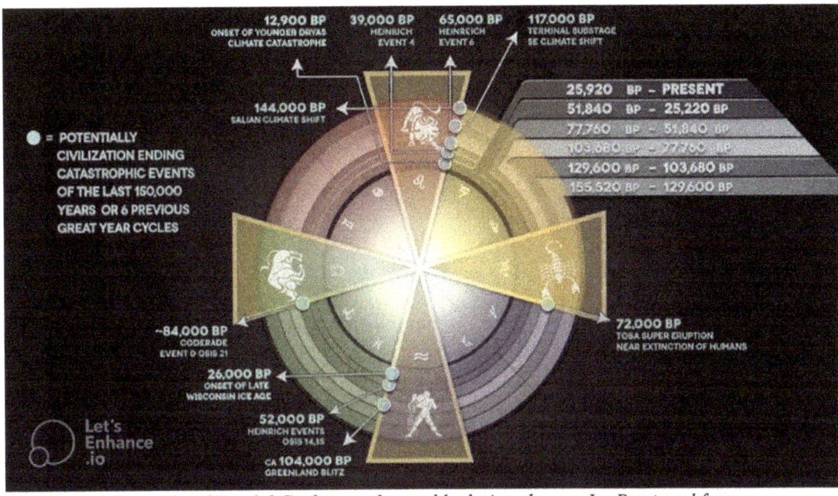

Courtesy of the works of Randal Carlson, enhanced by let's enhance. Io. Retrieved from https://sacredgeometryinternational.com/the-great-year-of-the-world-sacred-geometry-cosmic-cycles-and-catastrophe/

Another researcher who has a similar set of events for distinct reasons is the late, great Douglas B. Vogt. He is the founder of the Die-Hold foundation, and his timeline is extraordinarily similar to Carlson's and to Lacroix's as well. These various timelines fit rather nicely together as a package of evidence that is incredibly significant to all of humanity and our survivability at the coming end of the age. Vogt calls his findings a Clock Cycle:

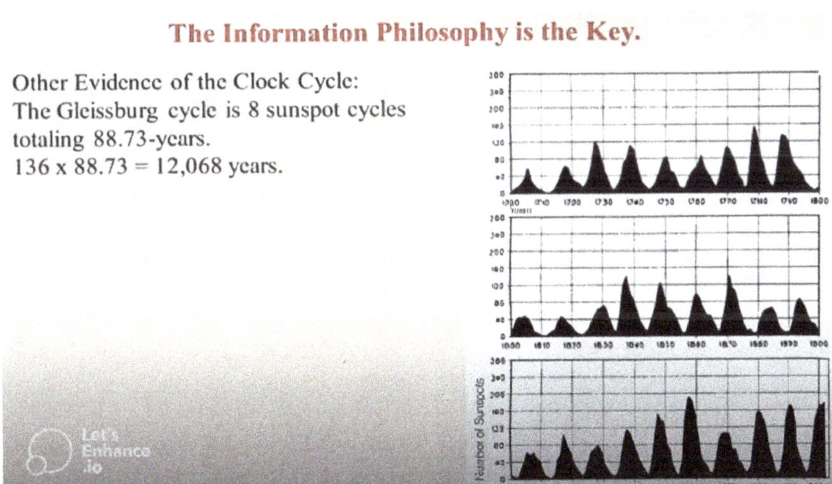

Courtesy of the works of Doug Vogt, and the die-hold foundation, enhanced by Let's enhance. Io. Retrieved from https://dieholdfoundation.com/

One must take note of the important synchronicity that is emerging around every twelve thousand years or so. There is a natural cosmic reset of sorts that has a profound effect on biological organisms on Earth at or near that very inflection point over time. We will get into more specifics later in the book; the major takeaway here is that there is something going on of a cyclical pattern that is not random. These patterns suggest that the next repeat of major events is upon us in the coming decades, and we must prepare now, before it is too late. Hopefully, you will see how these connections have meaning and purpose that the Creator of the universe is real, that the Creator of the universe is love, and God wants us all to find the divinely eternal. Mr. Vogt has also discovered blank periods that illustrate a solar clock cycle as follows:

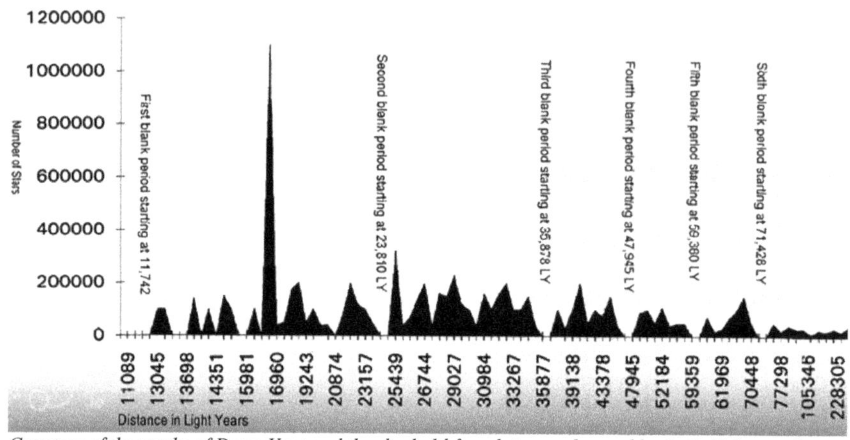

Courtesy of the works of Doug Vogt and the die-hold foundation, enhanced by Let's Enhance. Io. Retrieved from https://dieholdfoundation.com/

These blank periods are further explained in the next graphic, which yet again demonstrates a roughly twelve-thousand-year, repeating set of events that, when seen in totality, ought not to be ignored.

Reversal No.	Distance in LY	Stars Visable (LY)	Distance from last blank period
1	11,742	1,304	N/A
2	23,810	1,630	12,068
3	35,878	977	12,068
4	47,945	978	12,067
5	59,360	1,304	11,415
6	71,428	5,870	12,068

The 6 Blank Periods in Space.

Courtesy of the works of Doug Vogt and the die-hold foundation, enhanced by Let's Enhance. Io. Retrieved from https://dieholdfoundation.com/

Doug Vogt claims to have discovered the same 12,068-year cycle in the sunspot cycles. A sunspot cycle is 11.09 years; eight of them are called a Gleissberg cycle, equaling 88.732 years. 136 Gleissberg cycles total 12,068 years. According to Vogt, the sunspot cycle is a resynchronizing frequency, while the Gleissberg cycle is a synchronizing frequency, and it is synchronized to the main clock cycle.

Vogt's works explain what causes the geomagnetic reversals and why they correspond to the ice ages. He presents evidence from numerous science journals that show six reversals and ice ages that carbon 14 (C14) dates to a nearly 12,000-year repeating cycle. He claims to have discovered the clock cycle three separate ways, which is hard to disprove. The most compelling evidence between Vought, Carlson, The ECDO, and La Croix's perspectives is that there is a cycle of events, and clearly something significant occurs around every 12,000 years or so for all four theories.

Call it the great year or the clock cycle, these various theories all speak of the same repetitive actions that have caused, and that will cause existential events to occur that can no longer be dismissed or ignored. It is true that Carlson argues for an impact event and Vogt for a solar nova. It becomes even clearer with the lens of a new perspective stemming from the discussions found in the latter section of this book, titled The Seers. Specifically, the Lady of Bledsoe's accounts took him away and, at one point, showed him the Earth from space. Bledsoe then saw the sun growing bigger than the Earth and blowing a sizable portion of the Pacific Ocean out into space. Bledsoe has no idea if this was a past event or a future event that he was shown. What if all of them, the ECDO, LaCroix, Carlson, Dunning, Davidson, Bleadsoe, and Vogt, are all right at the same time? Bledsoe's vision was both a past solar event and showing him, or better yet, through him, showing all of us a future event when the Sun will again do this nova thing. Doug Vogt has compared previous solar incursion evidence from past events, and he predicted that the next dust shell from our solar nova will hit the Pacific Ocean.

Here is where Carlson is also correct in that the oceans blown into space, as shown to Bledsoe, are frozen and eventually fall back to Earth, becoming extinction-level impact events as well. When you add in all the other research on lost civilizations, ancient technologies and major extinction level events that are known, a very clear picture begins to emerge on what the Earth and mankind can expect in the coming decades as we approach the end of the next cycle when the Sun will nova or micronova, and the poles of the Earth will shift places magnetically and possibly physically as well.

Please do find the works of these great scientists and the others listed in the various introductions, as this very brief overview barely begins to touch on the nuances of what is to come. What is certain at this point is that these existential events will take place between the 2030s and the late 2040s. While date setting is not a particularly good idea, it is important to understand the gravity of this knowledge and how the coming events will affect all of us, whether we believe it or not.

In Part II, we have demonstrated a unified theory that Earth's history is shaped by recurring catastrophic events, often cosmic in origin, which are cyclical and predictable. A vision from Chris Bledsoe is mentioned, where the sun is seen growing larger and blowing a portion of the Pacific Ocean into space. These commonalities suggest this could be both a past and a future event. Part II then shows the links between several theories by suggesting that the frozen ocean water blown into space eventually falls back to Earth, causing extinction-level impact events. The main points of Part II are the following:

- *Cyclical Catastrophism:* The alternative that challenges the traditional gradualist view of history, proposing that the rise and fall of civilizations are shaped by cycles of catastrophic change.
- *Solar and Galactic Cycles:* Where catastrophic events are linked to cycles of the sun and the Milky Way galaxy. The cycles are determined by *"electromagnetic sheets"*. The sun is theorized to go *"NOVA"* or *"MicroNova"* about every 12,000 years.
- *Randall Carlson's Perspective:* Randall Carlson's work is cited to support the idea of geological catastrophism. His theories include a massive cosmic impact event around 11,600 to 13,000 years ago that caused the Younger Dryas climate shift, and immense mega floods as a result. This event is believed to have wiped out or set back previous advanced civilizations. Carlson's *"Great Year"* is presented as a timeline based on the precession of the equinoxes, dividing a 25,920-year cycle into four *"great seasons"*.
- *Doug Vogt's Perspective:* Doug Vogt's research also supports a cycle of events occurring approximately every 12,000 years, but he attributes them to a solar NOVA event.

- *A Unified Theory:* Part II attempts to reconcile the ideas of Carlson, Vogt, Dunning, La Croix, the EDCO, and Bledsoe, suggesting that they are all correct.

Upcoming Event: Part II concludes with the idea that these existential events are expected to take place between the 2030s and late 2040s, and that the end of the current cycle will involve a solar nova and a shift in the poles of the Earth that is magnetic and/or physical. There is mounting evidence that supports this idea, and the outstanding question for those who believe should be: *"What are we doing to prepare for what is coming?"* We will now take a deeper dive into some of the science that is foundational to solar events and options on scientific ways to consider all these ideas down to the quantum level of existence.

Chapter 7
Various Scientific Theories

As the discussion moves forward, it becomes important to consider how these patterns extend beyond individual theories and begin to connect across multiple systems. The relationships between solar activity, geological processes, and historical events suggest that these cycles may not act independently, but instead operate as part of a larger, interconnected process. The following section explores these connections in greater detail.

Ben Davidson:

Ben Davidson looks at the electric universe and how the galactic current sheet as a wave intersects with the plane of our solar system in a cyclical manner, and this is in part what causes both the solar micro nova and the pole shift. Combined with cosmic rays from the galaxy and beyond, the light, particles, and magnetic fields of the sun interact with various layers of the Earth. These interactions range from the outer magnetosphere down through the ionosphere and atmosphere, and even affect the crust, mantle, and perhaps the core. Ben Davidson, as an author, is a recognized expert in solar-terrestrial physics and the founder of Space Weather News.

Ben has a significant online presence, sharing insights on solar activity and its implications. Ben has published breakthrough papers in solar-terrestrial physics and is one of the leading space weather analysts. Ben is the creator of *"TheSuspiciousObservers"* YouTube channel, sharing all things sun, Earth, and science-related for over a decade. He is also the creator of *"The Mobile Observatory Project,"* which is an educational outreach project. His organization sponsors Observing The Frontier conferences, and Observer Ranch, a campground where families can learn about the coming solar events, how to prepare, and survival techniques. His book serves as a resource for meteorologists and the public, simplifying complex scientific concepts related to the sun and its effects on Earth. Here are some of his notable works:

Weatherman's Guide to the Sun: A comprehensive guide that simplifies complex solar science for a general audience.

Space Weather: The book discusses how space weather affects various aspects of life on Earth, including weather patterns, climate change, and technology.

Solar Cycles: It explains the solar cycles and their impact on Earth, detailing how solar activity influences terrestrial phenomena.

Human Health and Technology: The relationship between solar activity and human health, as well as its effects on technology, is explored.

The Next End of The World: The Rebirth of Catastrophism explores themes of catastrophism and its implications for understanding future events. These titles reflect Davidson's commitment to making scientific concepts accessible to a broader audience.

Space weather has become a recognized actor in weather, long-term climate change, seismicity, technological performance, and biology. In his third edition of the literature review, Davidson covers about five hundred of the most important studies describing the interactions of Earth and sun, including a new chapter on extreme solar activity and Earth's catastrophe cycle. Davidson discusses how the belief in catastrophism was prevalent before the mid-1800s, when gradualism became the dominant view in geology and biology. He draws parallels between historical catastrophic events and current scientific understanding. Davidson integrates recent research on solar activity, electromagnetic fields, and geological phenomena to support his claims. Davidson suggests that these catastrophic cycles occur approximately every 12,000 years, potentially leading to major shifts in Earth's environment. Davidson's work aims to provide a comprehensive understanding of how cosmic and solar phenomena can impact Earth. He offers a challenge to the notion that changes occur solely through gradual processes. His works serve as a call to reconsider the implications of these catastrophic cycles on our planet's future. A major reason that we do not know or accept these into current mainstream scientific thought is the idea that Earth's Civilization has been destroyed several times over throughout the eons of the life of the Earth.

Juan Moritz & Erich Von Däniken:

In June 1965, Argentine archaeologist Juan Moritz, during an accidental excavation in Ecuador, discovered an incredible marvel. Moritz found a secret entrance to a vast tunnel system, guarded by an indigenous Indian tribe. The entrance leads some 250 meters underground and is estimated to be over 4,000 kilometers long. Inside the tunnel, the walls are smooth and polished, with numerous exquisitely crafted rock archways and large doors. Several of the spacious chambers are as large as hangars for jet airliners. German author Erich Von Däniken entered this tunnel accompanied by Moritz himself. Even Von Däniken, who is known for his bold imagination, was stunned speechless and unequivocally considered it the world's largest engineering project at the time, as well as the world's greatest and most difficult puzzle to solve.

William J. Meister:

In the summer of 1968, American scientists William J. Meister discovered a fossil in Utah with two complete human footprints. This shoe print is about 26 cm long and 8.9 cm wide. The 1.5 cm depression in the heel suggests it was a shoe like modern casual footwear, meaning the owner lived in an environment with a certain level of civilization. Research indicates this fossil dates back 270 million years. Surprisingly, the footprint is also stepping on a trilobite, which is one of Earth's oldest creatures. Examination confirmed the fossil belongs to the Cambrian stratum, which is about five hundred million years old. In such a distant era, who could have left these footprints? This discovery could have not only challenged the theory of evolution, but it also might completely tear it down to an inconsistent pseudoscience at best. For this reason, the mainstream sciences have, quote, unquote, debunked him. This is a modern example of what happened to Copernicus, otherwise known as the Galileo effect.

Others:

Marine scientists from the Exploration Vessel Nautilus unexpectedly found a "roadway" on the seabed off the coast of Hawaii in the Pacific. The formation, which resembles a yellow brick road, was discovered in the Papahānaumokuākea Marine National Monument while the crew was investigating underwater volcanic activity. If you fitted a submarine with wheels, you could even drive it along this road like a bus.

August Dunning Insights: Dunning has critiqued Vogt's work and expanded upon it with similar ideas to Davidson concerning the electric universe and electromagnetic current sheets. These are considered a major driving force that causes solar micronova and planetary pole shifts in long-term cycles stretching over eons, ranging from tens of thousands of years to hundreds of millions of years. Dr. August Dunning's *"The Theory of Everything"* proposes a novel framework for understanding the universe, emphasizing the interplay between space and counterspace within a dielectric vacuum field. There are several points summarizing Dunning's theory.

The universe is a singular field with two primary modalities: Electromagnetism and magnetism, which are expressions of a fundamental dielectric vacuum field. Space and counterspace are essential aspects of this dielectric vacuum field. Counterspace represents vast, unexpressed inertial energy within this background vacuum field, while space manifests as perturbations of this field. Matter, including subatomic particles, is composed of accumulations of electrons and positrons, which are created from colliding photons.

These electrons and positrons possess inherent magnetic properties that drive their aggregation into the more complex structures observed as atoms and, eventually, matter. There are no *"particles"* in the traditional sense, but rather regions of perturbation within the dielectric field, manifested as force in motion. These perturbations are initiated by light collisions, creating regions with a Z-axis expression that is more commonly called dimension within the counterspace. Magnetic fields are expressions of the electric vacuum field, exhibiting volume and magnitude. Everything in the visible universe, from the smallest subatomic particles to galaxies, is fundamentally magnetic and interconnected through the underlying dielectric field.

The Space/Counterspace theory offers a mechanism for the creation and expansion of the universe, explaining how the electric vacuum can expand to express matter and energy. Dunning also uses this frame of reference for current sheets and solar interactions that cause micro nova in much the same way that Ben Davidson does. There are far more similarities between the two theories than there are differences. Because of this, there is no issue of who is correct, but rather how we consider them all as a valid approach to the question of the sun, nova, and pole shift events.

What Are Electromagnetic Current Sheets?

I am not an expert in this or any of these scientific fields of study. I am simply reporting what information I have gathered in a way that hopefully is easier to understand and demonstrates connections. As I have come to understand it, an electromagnetic current sheet on a galactic scale is a large, relatively thin region in interstellar space where the direction of the magnetic field reverses. At these boundaries, significant electrical currents flow between them. These structures are fundamental to the dynamics of magnetized plasma in space and are often associated with active galactic nuclei and cosmic ray acceleration, which are most often connected to black holes.

Electromagnetic Current Sheets, Key Characteristics:

Plasma Dominance: Space is filled with plasma or ionized gas, not empty space. The behavior of electromagnetic fields in this environment is governed by the principles of plasma physics.

Magnetic Field Reversal: The defining feature of a current sheet is the sharp transition where the magnetic field changes direction. This localized, thin boundary confines magnetic fields of opposing polarity.

Current Flow: The change in magnetic field across the sheet necessitates the flow of an electrical current, as dictated by Ampère's Law. These currents can be vast, spanning many light-years.

Energy Dynamics: Current sheets are sites where magnetic energy can be efficiently converted into kinetic energy and heat, often through a process called magnetic reconnection.

This is a powerful mechanism for accelerating charged particles, such as cosmic rays, to remarkably high energies.

Scale: Current sheets occur at various scales in space, from a more local sheet in the Earth's magnetotail, around the Sun, to a galactic-scale current sheet that can extend across vast distances. These structures play a crucial role in the transport of energy and matter within galaxies and are a key area of study in astrophysics and plasma physics within the galaxy's interstellar medium or within the powerful jets emanating from a central black hole. An electromagnetic current sheet on a solar system scale is a large, thin region in space where the direction of the magnetic field reverses, causing an electrical current to flow across the sheet.

Structure: The current sheet is a boundary separating regions of space with oppositely directed magnetic fields. It is relatively thin compared to its vast extent, often resembling a warped surface or disk.

Location: In our solar system, the most prominent example is the heliospheric current sheet, which separates the two hemispheres of the Sun's magnetic field. The Sun's rotation causes this sheet to become warped into a *"ballerina skirt"* shape.

Formation: Current sheets form naturally in plasmas where magnetic fields are present. The magnetic field lines from one region connect to those from the other across this boundary.

Significance: Electromagnetic current sheets play a crucial role in space weather and the dynamics of the solar wind. Charged particles interact with the current sheet, and events like magnetic reconnection can occur here, releasing substantial amounts of energy and affecting the space environment. The following illustration is an attempt to provide a visualization of our solar magnetic current sheet.

Courtesy: https://fineartamerica.com/featured/1-heliospheric-current-sheet-mikkel-juul-jensenscience-photo-library.html

In Chapter 7, we discussed the connections between electromagnetic current sheets and their cycles over large scales of distance and time that are a major contributor to solar nova / micro nova. The theories of Ben Davidson, August Dunning, and others discussed within Part II of this book relate to the science behind large-scale cosmic cycles, showing the connection to solar nova / micro nova through the concept of the galactic current sheet. The core idea of a Galactic Current Sheet Interaction is that the solar system periodically intersects with the galactic current sheet, which is a vast, thin region in interstellar space where the magnetic field polarity reverses. Some common ideas that flow among these theories are as follows:

Triggering Event: This interaction introduces dust and energetic particles into the Sun's atmosphere. The buildup of this material, combined with the Sun's magnetic field dynamics, eventually triggers a solar micro nova event.

Cyclical Galactical Solar Bouleversement: This term is defined by an interaction that is proposed to be cyclical, occurring approximately every 12,000 years. This cycle is presented as a primary driver of historical catastrophic events and major shifts in Earth's environment, including climate change, solar micro novae, and magnetic pole shifts.

Effects on Earth: The resulting solar micro novae and associated space weather phenomena are suggested to have profound impacts on Earth's magnetosphere, atmosphere, crust, and even potentially the core, leading to the *"catastrophe cycle"* referenced by many of the authors listed in the introduction to Part II. This framework challenges mainstream scientific views by suggesting a cyclical, external galactic mechanism as a primary driver of major, rapid Earth changes, in contrast to the more gradual processes typically accepted in geology and biology. Chapter 6 looked closely at recent research and theories, especially those presented by Ben Davidson and August Dunning, which have significantly advanced our understanding of the connections between galactic and solar cycles, with a particular focus on electromagnetic current sheets and their role in triggering solar micro nova events and pole shifts. These cycles, supported by a growing body of scientific literature and observations, suggest that the solar system periodically interacts with the galactic current sheet. The galactic current sheet is a vast, thin region in interstellar space where the magnetic field reverses direction. This interaction introduces dust and energetic particles into the Sun's atmosphere, which, when combined with solar magnetic dynamics, can trigger powerful solar micro nova events.

Chapter 7 builds upon Chapter 6 to show that cyclical interaction between the solar system and the galactic current sheet is proposed to be a primary driver of historical catastrophic events, including climate change, solar micro nova, and magnetic pole shifts. These cycles challenge conventional gradualist models in geology and biology, suggesting that rapid, external galactic mechanisms may be responsible for major shifts in Earth's environment. The resulting effects on Earth can be profound, impacting on the magnetosphere, atmosphere, crust, and possibly even the core, contributing to what Davidson and others refer to as the *"catastrophe cycle."*

The connections outlined by Davidson, Dunning, and other researchers encourage a reconsideration of how cosmic phenomena influence Earth's history and future. Their work synthesizes hundreds of studies on solar-terrestrial interactions, extreme solar activity, and geological catastrophes. While mainstream science remains cautious, the accumulating evidence and theoretical advancements provide compelling support for the significance of galactic and solar cycles, electromagnetic current sheets, and their potential to drive major, rapid changes on the Earth.

Chapter 8
Solar Evidence

In examining solar activity, it is important to consider not only theoretical models but also observations that are already recognized within mainstream science. Many of these observable phenomena provide insight into how the Sun behaves during periods of heightened activity and how such events may affect surrounding bodies. The following section presents a brief overview of these observations and their implications.

Solar Flares and Nova are Reddish in Appearance: Solar flares and Novae are massive explosions on the sun's surface that release bursts of radiation, including visible and ultraviolet (UV) light. When a strong solar flare erupts, it can significantly increase the brightness of the sun as seen from the Moon and from Earth.

Radiation and Light Emission: While the primary light from the sun is white, solar prominences, which are often associated with flares, and now novae that emit light in specific wavelengths, including the Hα line, which is a red to orange coloration. The overall increased brightness during a flare or nova would be dominated by a wider spectrum of light, including UV and X-rays.

The Moon's Lack of Atmosphere: The Moon has almost no atmosphere. This means there is no atmospheric scattering of light like *"Rayleigh scattering,"* which makes Earth's sky blue, sunsets red, and effects like the *"blood moon"* effect seen during a lunar eclipse, where the Moon is in Earth's shadow. Instead, the Moon's surface would be directly exposed to the increased radiation from a large solar flare or nova event.

Space Weathering and Sputtering: Solar flares and novae are sometimes called coronal mass ejections (CMEs), which can impact the Moon's surface, a process known as space weathering. These high-energy particles can cause sputtering, where they collide with and eject material from the Moon's surface.

During a strong CME, a massive cloud of plasma can act like a *"sandblaster"* on the lunar surface. It is possible that the ejected material or changes in the surface composition could, under certain conditions, alter the reflectivity of the Moon's surface, potentially leading to a reddish or orange hue.

Specific Flare Intensity: While it is difficult to quantify the precise intensity of a flare or nova needed, they would need to be strong enough to cause significant space weathering effects on the lunar surface to produce a noticeable color change. The most powerful solar flares are classified as X-class flares, while novae have yet to be classified; they are exponentially stronger than X-class CMEs. Current X-class rated CMEs are known to cause a range of effects on Earth, including radio blackouts and radiation storms. On the Moon, without the protection of an atmosphere or magnetosphere, the effects of such flares would be even more pronounced.

A solar nova would be even more devastating for both the Earth and the Moon. It is important to note that the *"red glow"* from a solar flare impacting the Moon is a different phenomenon from a *"Blood Moon."* The former is caused by the direct impact of energetic particles and radiation on the lunar surface, potentially altering its reflectivity. The latter is an optical effect caused by Earth's atmosphere during a lunar eclipse. A strong outburst from the sun, particularly a powerful solar flare or nova, can make the Moon appear to glow red.

Energetic Processes, Natural Formation: In space, energetic particles, and shockwaves impact precursor materials like silicon carbide dust, leading to fullerene formation. Fullerene formations are more commonly called buckyballs. Lightning strikes, a phenomenon characterized by extremely elevated temperatures and rapid energy release, are also associated with buckyball formation. The formation of buckyballs hinges on providing enough energy to break down carbon-containing materials into individual atoms or small clusters. The breakdown is then followed by rapid cooling and controlled conditions that favor the formation of the distinctive fullerene formations. In environments like the interstellar medium, where fullerenes are observed, energetic particles and shock waves can significantly impact precursor materials, such as silicon carbide dust, leading to the formation of fullerenes or buckyballs.

Fullerene (Buckyball) Formation:

Ultraviolet (UV) Radiation and Photo-processing: Strong UV radiation fields, common around massive stars and in planetary nebulae, can interact with carbonaceous materials like polycyclic aromatic hydrocarbons (PAHs) and hydrogenated amorphous carbon (HAC) nanoparticles. UV photons can cause dehydrogenation and structural transformations within these precursors, leading to the formation of pentagonal rings and ultimately the closed-cage structure of fullerenes or Buckyballs. This *"top-down"* approach, where larger molecules are broken down and restructured, is considered a significant pathway for buckyball formation in space. The continuous heating from UV absorption can keep particles in a vibrationally excited state, facilitating the structural transformations required for buckyball formation.

Shock Waves: Like those generated by supernovae, a nova shock wave can strongly alter the particle phase of carbonaceous materials, including silicon carbide dust. These shock waves can lead to the fragmentation and rearrangement of precursor molecules, creating conditions favorable for buckyball formation in the gas phase. However, the low post-shock carbon densities in space can hinder buckyball formation through vaporization/combustion routes alone, according to IOPscience.

Energetic Particle Bombardment (e.g., Cosmic Rays, Electron Impacts): High-energy cosmic rays, electron impacts, and even impacts from fullerene molecules themselves can interact with precursor materials and initiate processes leading to buckyball formation. Studies have shown that impact events on graphite surfaces can lead to the disintegration of impacting fullerenes and the ejection of carbon clusters and chains from the surface, creating potential building blocks for buckyball formation.

The *"knockout"* of individual carbon atoms from larger PAH species by energetic particles can effectively create the necessary pentagonal rings and lead to the formation of fullerene structures or buckyballs.

The Role of Silicon Carbide Dust: Silicon carbide (SiC) dust is a common precursor material found in space environments and has been experimentally demonstrated to influence buckyball formation. Laboratory studies have shown that incorporating silicon carbide additives into graphite electrodes can increase the yield of fullerenes in the produced soot during arc discharge synthesis. This suggests that the presence of silicon atoms in the reaction zone, along with increased arc power for evaporation, can enhance the efficiency of buckyball formation. Silicon carbide's role might involve providing a template or catalytic surface for carbon atom arrangement or altering the local conditions (e.g., temperature, density) in the region of buckyball formation.

Fullerene (Buckyball) Formation: The processes described for buckyball formation include UV radiation, shock waves, energetic particles, and silicon carbide dust, which all occur throughout the interstellar medium and in the environments of various stars. These are projected out into space from the star during novae events, which is why we see them on Earth, the moon, and even on Mars.

Megalithic Statues: The presence of melted megalithic statues is an intriguing archaeological observation. Some hypotheses involve the use of tools, chemicals, or even vitrification in high heat, but these are speculative and require further investigation. The high heat can only come from a few sources: a significant impact from space, a nova event, or a thermonuclear explosion.

When the next nova/micronova occurs, the sun will blow outward a spherical dust shell that will reach the Earth within 17 to 18 hours. Some of the materials found on the Earth and on the moon are little, tiny glass beads. Some of the beads also have fission tracks on their surface from aluminum twenty-six or barium ten. There is also a whole bunch of radioactive elements that can only come from relativistic temperatures from atomic bombs going off, or from a solar nova event. When the dust shell hits first, the water gets evaporated. We are going to have a bunch of clouds around the Earth with very thick clouds, because a lot of water has evaporated. The water will go up in the form of steam high into the atmosphere and will be very dense. The Pacific Ocean will be a boiling mess when that dust shell hits. It is going to blow a lot of the ocean and some land away on the impact side of the Earth.

According to Doug Vogt, *"The last time the Sun hit, I believe along this line here roughly either 105 or 110 degrees longitude, where China is and India." "Because China and India saw the dust shell impact the last time, and we have evidence of bone with fission tracks in it."* So, the next dust shell impact will be that many degrees east out into the Pacific Ocean during the next solar nova.

Chapter 9
The Scientific is Prophetic

There have been several scientists and theorists who have contributed to connecting things like relativity and quantum mechanics. In the totality of the information that this book brings together, connections are a paramount consideration. Let us now bring some of those contributors to light on the pages of Chapter 9 as these theories impact the information covered in previous chapters.

Who Is Hannes Alfvén?

Hannes Alfvén (1908–1995): A Swedish electrical engineer and plasma physicist renowned for his pioneering work in magnetohydrodynamics (MHD) and plasma physics. He was awarded the Nobel Prize in Physics in 1970 for his fundamental contributions to the understanding of magnetized plasmas, which are essential for explaining phenomena in space and astrophysics. Alfvén introduced the concept of Alfvén waves, or transverse waves that propagate along magnetic field lines in plasma and was a leading advocate for the Plasma Universe theory, which emphasizes the importance of electromagnetic forces in shaping cosmic structures. His work laid the foundation for modern space science, influencing everything from solar physics and auroras to the behavior of interstellar matter.

Who Is Immanuel Velikovsky?

Immanuel Velikovsky (1895–1979): A Russian-born psychiatrist, psychoanalyst, and author best known for his controversial books proposing radical reconstructions of ancient history and astronomy. Velikovsky challenged mainstream scientific and historical consensus, suggesting that catastrophic celestial events shaped human civilization and that ancient texts and myths contain records of these events. His most notable theory, presented in Worlds in Collision, posited that planetary disasters, particularly involving Venus and Mars, had dramatic effects on Earth and were witnessed by ancient peoples. Despite criticism from the scientific establishment, Velikovsky's work influenced discussions in fields ranging from mythology and archaeology to astronomy.

Who Is Edgar Cayce?

Edgar Cayce (1877–1945): Often referred to as the *"Sleeping Prophet,"* Edgar Cayce was an American clairvoyant who gained fame for his trance readings on topics ranging from health and holistic healing to ancient civilizations and prophecy. During self-induced sleep states, Cayce provided thousands of psychic readings for individuals, offering guidance on physical ailments, spiritual growth, and the mysteries of human existence. His work laid the foundation for the field of holistic health and inspired ongoing research into consciousness, spirituality, and alternative medicine. The Edgar Cayce readings continue to influence both popular and scholarly interest in mysticism, reincarnation, and the intersection of science and spirituality.

More on Doug Vogt:

Douglas B. Vogt (1947–2023): Author of *"The Theory of Multidimensional Reality,"* which is the world's first purely information theory of existence. Vogt's book suggests that matter is not the dominant thing in our Universe, but rather information creates our matter world. His book is the culmination of 46 years of study, resulting in logical explanations for the hardest phenomena known. Mr. Vogt first presented his theory in his first book, Reality Revealed, The Theory of Multidimensional Reality, in 1977.

He further refined the theory in the 2007 book, *"God's Day of Judgment, the Real Cause of Global Warming."* This book uses Vogt's theory to explain what causes the geomagnetic reversals, or polar reversals, and why they precede the ice ages. Mr. Vogt proposes in his works the exact number of years between reversals explained by the discovery of a clock cycle in the universe that causes geomagnetic reversals. Vogt's theory can explain the most difficult phenomena using this same philosophic framework. His theory unites parts of quantum mechanics with parts of string theory. His theory does this by first assuming that the universe is the product of information and that the information forms the matter world we live in. As reported on his website, the die-hold foundation, Mr. Vogt presents *"The Theory of Multidimensional Reality,"* which explores the nature of existence through an information theory lens. His ideas address complex questions about reality, aiming to provide clarity on difficult scientific concepts. His works are intended for readers interested in the intersection of philosophy and science, particularly those curious about the underlying principles of reality and existence.

In 2015, Vogt published *"The Theory of Multidimensional Reality,"* where he proposes that existence and the physical universe can be best understood through the framework of information theory. The Vogt perspective can shed light on fundamental physical phenomena such as light, gravity, time, and Planck's constant by interpreting these core aspects of physics through an informational lens. Vogt seeks to demystify complex questions about the nature of reality. This approach aims to bridge the gap between philosophy and science, offering readers new insights into the foundational principles that govern existence. Vogt's theory is particularly suited for those with a keen interest in both theoretical physics and philosophical inquiry, encouraging a deeper exploration of the informational structure underlying the universe.

According to Vogt's theory, *"The information that matters comes into this dimension along the y axis, which is 90 degrees out-of-phase from the matter world, the z axis. The following graph depicts this idea. Time can be simply explained as the rate at which information is modulated into our existence."*

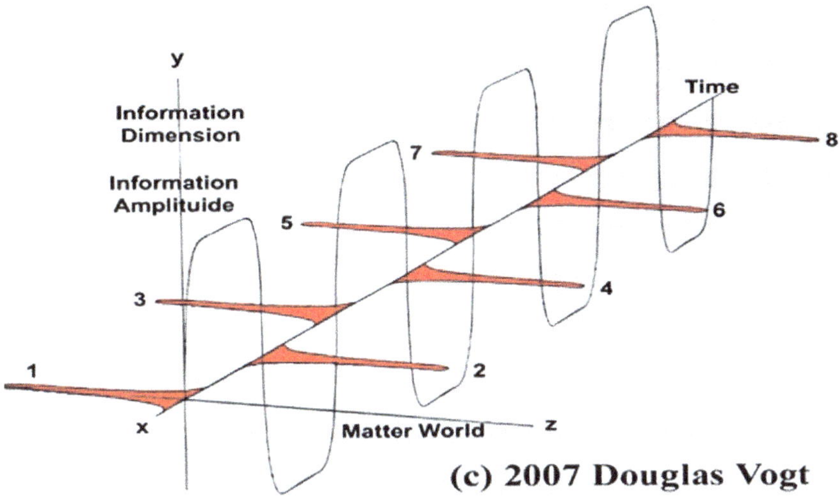

Courtesy of the works of Doug Vogt, and the die-hold foundation, enhanced by Let's enhance. Io. Retrieved from https://dieholdfoundation.com/

He then connects to string theory, suggesting that it assumes that a one-dimensional string vibrates at a specific frequency and traces out a plane or vibrating ring shown above. Vogt suggests that "each frequency creates a different subatomic particle…," and "Closed-end string theory attaches two opposite sides to form a tube." Vogt's theory takes the tube and connects the two open ends together to form a toroid. Vogt then goes deeper with: "The theory holds that hundreds or thousands of separate frequencies form each atom. It is transmitted on a carrier wave that pulses at a common frequency, which we will assume is Planck's time." Vogt's ideas are substantiated by an Italian team at EPFL; EPFL stands for École Polytechnique Fédérale de Lausanne, which is a renowned research university located in Lausanne, Switzerland. The EPFL is recognized for its excellence in engineering, technology, and the natural sciences, and is one of the two Swiss Federal Institutes of Technology.

EPFL conducts innovative research and offers education at the undergraduate, graduate, and doctoral levels, attracting students and researchers from around the world. Led by Fabrizio Carbone, researchers were able to capture the first-ever snapshot of the dual behavior of light. Notice that regardless of the color frequency change, they all pulse at the same rate. Vogt suggests that his *"Theory of Multidimensional Reality successfully explains what light is and why it possesses both wave and particle properties."* Doug then suggests that his theory and the evidence presented in his works prove that the reality we live in is the product of a synchronistic system that has clock cycles in it. He goes on using the German gravitational wave detector observatory as more evidence. Called GEO600, who concluded that the detected *"noise"* could not be explained until Juan Maldacena proposed the idea that the *"noise"* was the result of the possibility that our Universe was a hologram.

Who is Juan Maldacena?

Juan Maldacena (born 1968): An Argentine American theoretical physicist best known for formulating the AdS/CFT correspondence, a groundbreaking concept in string theory and quantum gravity. This duality, also called the Maldacena conjecture, proposes a deep relationship between certain theories of gravity defined in higher-dimensional spacetime or Anti-de Sitter space and quantum field theories defined on the lower-dimensional boundary of that space. Maldacena's work has had a major impact on theoretical physics, providing new insights into black holes, quantum entanglement, and the nature of spacetime itself. He is a professor at the Institute for Advanced Study in Princeton, New Jersey, and is recognized as one of the leading figures in modern physics for his influential research contributions.

Who is Christopher Langan?

Christopher Langan (born 1952): An American autodidact. An autodidact is a person who is self-taught, acquiring knowledge and skills through independent study rather than formal education or traditional schooling. Autodidacts typically pursue learning driven by curiosity and personal interest, often mastering complex subjects on their own initiative.

Mr. Langan has been described in popular media as one of the smartest individuals in the United States, with a reported IQ score that exceeds 190. He is known for his exceptional intellectual abilities. He is best known for proposing the *"Cognitive-Theoretic Model of the Universe"* (CTMU). CTMU is a comprehensive philosophical framework that attempts to explain the nature of reality, consciousness, and the relationship between mind and universe. Langan's ideas, which blend logic, metaphysics, and information theory, have attracted both interest and debate within intellectual circles. Despite limited formal academic involvement, Christopher Langan continues to contribute to discussions on philosophy, science, and the foundations of existence. His CTMU theory aims to provide a comprehensive description of reality, integrating concepts of cognition and the universe. Langan describes the CTMU as a *"Self-Configuring Self-Processing Language"* (SCSPL), which posits that reality can be understood as a language that communicates with itself. Some of Langan's concepts are as follows:

Self-Configuring Self-Processing Language (SCSPL): This concept suggests that the universe operates as a self-referential system, where the laws of reality are generated and processed internally.

Telic Recursion: This is a two-stage process in which the primary stage establishes the fundamental laws of physics, while the secondary stage creates specific adaptations as reality evolves.

Cognition and Reality: Langan argues that cognition and the universe are interconnected, asserting that understanding this relationship is crucial for grasping the nature of existence.

The CTMU claims to bridge the gap between mind and reality, suggesting that understanding the universe requires a cognitive approach. Langan believes that this model can address fundamental questions about existence, purpose, and the nature of reality itself. The Cognitive-Theoretic Model of the Universe (CTMU) proposes a unified framework where mind and reality are deeply interconnected and self-configuring. This theory aims to bridge the gap between science, philosophy, and metaphysics. Some of the vocabulary that is required for understanding the CTMU is as follows:

1. *Self-Configuring, Self-Processing Language:* Is just an intrinsic, completely self-contained language that is comprehensively and coherently self-descriptive and can thus be model-theoretically identified as its own universe or referent domain.
2. *Telesis:* Progress that is intelligently planned and directed.
3. *Infocognition; "Self-transducing information":* where information and cognition are recursively interdefined, the nexus point is infocognition.
4. *Unbound telesis:* Is the ground state of existence arrived at by stripping away the constraints of reality.
5. *3 Ms:* MAP (closure) M=R (Comprehensiveness) MU (consistency).
6. *Topological Descriptive Duality:* A topology serves as a dual or opposite of its corresponding description.
7. *Constructive filtrative Duality:* Simply asserts the general equivalence of these two kinds of process with respect to logico-geometric reality.
8. *Conspansive Duality:* Suggests that there is no difference between the expansion of the universe with respect to its contents, and the contraction of its contents with respect to it.

9. *Triality:* A relationship between three vector spaces.
10. *Conspansion:* Refers to a dual process describing how reality evolves. Specifically, it suggests there is no essential difference between the expansion of the universe relative to its contents and the contraction of its contents relative to the universe itself. This concept highlights a reciprocal relationship, where both the growth of the universe and the compression or internalization of its contents are two sides of the same underlying process within the CTMU framework.
11. *Inner Expansion:* Generative phase of conspansion.
12. *Requantization:* Is a process of conspansive that changes Planck's constant by applying an internal scaling factor to the nesting.
13. *Extended Superposition Principle:* Principle based on conspansive duality, which puts separated events of space and time in immediate contact with each other.
14. *Sums Over Futures (Interpretation of Quantum Mechanics):* Quantum phenomena are not random but obey higher-order causal invariants.
15. *Universe as a self-representation entity:* Four-dimensional curved spacetime of general relativity.
16. *Telic Principle:* That reality must select itself for existence.
17. *Generalized Utility:* The quantity that reality exists to maximize.
18. *Telic Recursion:* The process by which universal free will exists.
19. *Telor:* Syntactor creating a telon.
20. *Telon:* Is a potential state of high utility for one or more telors.

21. *Teleological Evolution:* A process of alternating replication and selection through the universe creates itself along with the life it contains.
22. *Teleological Consistency:* In the CTMU, teleological consistency refers to the requirement that the universe's evolution and events remain logically and purposefully aligned with its own self-selection and maximization of utility. This means that the unfolding of reality is not arbitrary but coherently directed toward the fulfillment of its intrinsic goals, ensuring that all actions and developments support the universe's continued existence and purpose.
23. *Principle of linguistic Reducibility:* Reality is a self-contained form of language and fits the definition of language.
24. *Syntactic Operator:* Self-traducing information.
25. *Syntactic Covering:* A descriptive containment of one syntax by another.
26. *Unisection* - The process of finding what things have in common.
27. *Distributed Syntax:* Universal laws of nature.
28. *Metasyntax:* Describes the allowable structure and composition of phrases and sentences of a metalanguage.
29. *Human Cognitive Syntax:* A set of structural and functional rules obeyed by the human mind.
30. *Distributed Solipsism:* A type of solipsism where oneself is distributed over all individuals.
31. *Generalized Cognition:* The aspects of the mind and mental functioning that reality itself uses to evolve.
32. *Hology:* A logical analog of holography characterizing the most general relationship between reality and its content.

33. *Holotheism:* A theological system implied by logical theology.
34. *Self-Determinacy:* An object that determines its own cause from within itself.
35. *Self-Multiplexing:* The generation of multiple possible ways for reality to evolve itself, within reality itself.
36. *Self-Selection:* Reality having only itself to create itself out of and only itself to describe its own purpose in relation to.
37. *Supertautology:* A theory of reality constructed in a way that guarantees that it is true.
38. *Syndiffenesis:* The expression and/or existence of any difference relation entails a common medium and syntax.
39. *Noeon:* A quantum of infocognition.
40. *Closed Descriptive Manifold:* Refers to a structure or framework that serves as a self-contained system for describing reality. This concept is central to understanding how reality is modeled as a unified, self-descriptive logical structure. The term suggests that all possible descriptions or states within this manifold are internally related, and the system does not rely on anything external for its definition or operation. The manifold is akin to a mathematical or logical space of descriptions that is *"closed"* in the sense that every aspect or element needed to define reality exists within the system itself.

This means reality, as described by the CTMU, is self-referential, self-contained, and its descriptive capacity is not dependent on anything outside itself.

41. *Protocomputation:* Refers to a fundamental or primordial form of computation that underlies reality itself. Unlike standard computation, which involves the manipulation of

symbols or data within a predefined system, protocomputation is thought to operate at the most basic level of existence, forming the very basis for logical structure, causality, and information processing in the universe.

In CTMU theory, protocomputation is not limited to physical or digital systems but is instead an abstract process that enables the self-organization and evolution of the universe's structure and dynamics. The CTMU suggests a unifying principle or structure underlying these phenomena, consistent with the synthesis arguing for a continuous interaction with the divine. According to Langan's theory, the universe is a *"self-referential system capable of self-configuration"* where reality and consciousness are interconnected. Where a seer uses free will to say yes to extra-dimensional interactions with the divine must be reinterpreted not as external intervention but as integral to the ongoing self-configuration. This self-configuration is a cognitive process of reality itself. Traditional notions of divine intervention must be reassessed, potentially viewing them as a *"teleo-grammatic functionality"* where the divine influences the evolution of reality through *"syntactic self-distribution." This process is* otherwise known as omnipresence, and *"perfect autotransductive reflexivity"*, otherwise known as omniscience, and *"self-configuration up to freedom"*, otherwise known as omnipotence.

Shifting of Paradigms:

The CTMU is a logical and mathematical framework for a theory of everything. The CTMU offers a means to bridge the gap between scientific and religious paradigms. Proper integration will lead to a new understanding of reality, potentially supporting a change in basic assumptions suggested in the introduction at the beginning of this book and tied into each section thereafter.

In essence, applying the CTMU in the act of synthesizing a *"Theory of Everything"* allows us to explore whether the totality of information provided in this work can be understood as a fundamental aspect of the universe's self-organizing and self-cognitive nature, as described by the CTMU. This, in turn, gives the reader a new way to consider and exercise their own free will to realize a more eternal perspective and what will be required to make it through the coming Earth changes. Chapter 9 uses Douglas Vogt's *"The Theory of Multidimensional Reality"* as a unifying framework to bridge seemingly disparate ideas across science, alternative history, spirituality, and theoretical physics. The connections are drawn together to support the concept of a self-organizing, cognitive reality that involves continuous interaction with a divine intelligence that is tied together in Christopher Langan's CTMU.

Key Connections:

Science and Alternative History: Chapter 9 connects established scientific figures with more controversial ones to suggest a broader understanding of reality.

Hannes Alfvén, a plasma physicist, and Immanuel Velikovsky, a catastrophic historian: Implicitly linked by their focus on the importance of electromagnetic forces in shaping the cosmos and potentially Earth's past, contrasting with or expanding upon mainstream geological timelines.

Science and Philosophy: Douglas Vogt's theory itself unites elements of quantum mechanics and string theory by proposing that information creates our material world, bridging the gap between physical science and philosophical/metaphysical inquiry.

Mind and Reality: Christopher Langan's *"Cognitive-Theoretic Model of the Universe"* (CTMU) is introduced as a comprehensive philosophical framework that argues for an inherent link between cognition or the mind, and the universe's self-configuring nature or what we experience as reality. This provides the underlying logic for the other connections.

Convergences:

There is an intriguing opportunity to identify shared themes and conceptual correlations between these seemingly disparate ideas, and Langan's *"Cognitive-Theoretic Model of the Universe"* (CTMU) suggests a potential avenue for their integration into a unified theory of everything. The inherent human drive to understand existence has led to diverse models of reality, from religious to scientific and philosophical. The Cognitive-Theoretic Model of the Universe (CTMU) is a complex philosophical system proposed by Christopher Langan that aims to explain reality through a self-referential, cognitive-theoretic lens. Despite their differing origins and methodologies, exploring the similarities and correlations between these approaches may reveal a deeper, unified understanding of existence. The CTMU is one of the most difficult parts of this text to grasp. How can all these scientific ideas be internalized, understood, and incorporated into the everyday lives of the average person?

Unifying Theme:

The overarching connection found throughout Part II is the proposition that reality is a self-referential, information-based system that operates with *"clock cycles"* and *"teleological consistency"* or purposeful alignment. This framework allows for the integration of physical laws with concepts typically considered spiritual or supernatural, suggesting a universe that is both physical and cognitive, and in continuous interaction with the divine. The ultimate goal is to offer a *"theory of everything"* that leads to a new understanding of reality, and a paradigm shift in human understanding and behavior that will allow humanity to overcome in love and live through the end of the age, into the next age, and on to eternity.

Chapter 10
Preparatory Considerations

Chapter 10 introduces a shift in perspective that challenges how prophetic warnings have traditionally been understood. Rather than viewing these messages as references to human-made destruction, this section explores the possibility that they point toward natural, large-scale cosmic events capable of reshaping life on Earth. Among these ideas is the concept of a solar micronova, a phenomenon still debated yet increasingly considered in relation to cyclical planetary changes and the possibility of a split in our timeline. When viewed through this lens, historical signs, and apparitions to be discussed may no longer represent symbolic warnings alone, but instead reflect patterns tied to recurring cosmic processes that carry profound implications for humanity and its future. We will get into more about the two timelines later in the book. Let us imagine we are here at some future point when the sun is going to go nova. What are the practical things that can be done to maximize survivability for you and for me?

Vogt and Davidson's Insights:

Ben is most concerned about the pole shift and that it will tilt the Earth's axis by 90 degrees. He feels that the only safe places are 5,000 feet or higher above sea level. Both Davidson and Vogt suggest high altitudes to avoid the sloshing seas that will be mega tsunamis that are several thousand feet high and go hundreds of miles inland. Davidson has named the Rocky Mountain area of Colorado as the new Valley of the Sun for this reason.

It is clear that high and underground, into non-volcanic mountains, are going to be the only safe places for a time in the near future for those who live through the solar nova. For Vogt, these safe zones will need to be small communities that work together for safety, food, skills, and other vital resources that will be required for survival. There will be a huge need for zero-point energy because the sun will not radiate bright enough to power solar cells for several years after the nova. The optimal approach is groups of families planning together who have various skills and are willing to make sacrifices for the good and survival of the group. These family groups must also be willing to help everyone so that as many as possible might survive.

The major focus for Vogt is to go underground with a sealed door at the entrance that will be able to withstand several miles of water, snow, and ice, as it will pile up miles high. He also says that the farther you are from the equator, the longer it will take for the ice sheets to melt over top of you. The reality is that there is no straightforward way to prepare or survive the events to come for any of us in most circumstances. Maybe the best that one can do is focus upon the eternal so that when we die, we are prepared for the next journey, and hopefully that means heaven.

Vogt's Sealed Cave Door Dilemma:

A custom forty by 25-foot watertight vault door designed to withstand over a million pounds of pressure would be a massive, highly specialized, custom-engineered product, likely costing millions of dollars. Off-the-shelf commercial or bank vault doors with standard security ratings generally cost from tens of thousands to hundreds of thousands of dollars, but do not meet the extreme specifications that a solar nova, pole shift, and ensuing ice age would demand.

Cost Drivers and Considerations: The cost of such a door is determined by extreme engineering requirements, materials, and size, far exceeding those of typical high-security applications.

Custom Engineering: A door of this size and pressure requirement falls outside standard production. It would require bespoke engineering, likely involving fluid dynamics and structural analysis, like those used in submarine hatches or high-pressure industrial vessels.

Material Strength and Thickness: To withstand one million pounds of pressure, the door and its frame would need to be constructed of incredibly thick, high-grade steel alloys. Another possibility is a composite of steel and specialized, high-strength concrete. Standard vault doors use steel plates of ¼ to one inch in thickness, which are inadequate for the pressure specified.

Size and Weight: At 40 by twenty-five feet, you have a door that is 1,000 square feet. The sheer volume of material needed would be enormous. This increases manufacturing complexity, shipping, and installation costs significantly.

Watertight Seal: The "watertight" requirement, combined with high pressure, necessitates a precision-engineered sealing system that can maintain integrity under extreme load, a complex and expensive feature. Integrating a reliable, high-pressure watertight seal into a door of this size and strength adds another layer of complexity and engineering cost.

Installation: Due to the enormous size and weight, installation would require specialized heavy-duty equipment, potentially costing a significant amount on its own.

Budget Estimation: While standard bank vaults can cost upwards of $150,000, and military-grade blast-resistant doors can range up to $2,000,000, a door with the outlined specifics entails extreme engineering demands that would be a one-off project whose cost would be determined by a specialist manufacturer after detailed consultation. The price would certainly run into millions of dollars. For an accurate quote, you would need to consult with a specialized high-security manufacturer or an engineering firm that manages extreme pressure applications. The price is driven by the extreme and highly specialized engineering requirements as follows:

Massive Dimensions: Standard and even most custom vault doors are designed for residential or commercial use with much smaller dimensions. Standard dimensions are typically up to eight feet high and seven feet wide. A 40x25-foot door requires unprecedented custom engineering and fabrication, moving it from a standard product to a one-off construction project.

Extreme Pressure Resistance: The requirement to withstand over a million pounds of pressure translates to significant engineering challenges. To put this in perspective, a specified door to withstand just three pounds per square inch (PSI) can handle over 9,000 pounds of total pressure across its area. A million-pound requirement would necessitate incredibly thick, specialized materials and a complex structural design, likely involving high-grade, thick-plate steel and robust internal reinforcement far beyond typical security applications.

Custom Engineering: This project would require a team of structural engineers, material scientists, and specialized manufacturers to design and evaluate the door. Engineering and architectural costs alone would be substantial.

Material and Fabrication: The sheer volume of high-strength, custom-fabricated materials are likely specialized steel alloys that would be enormous.

Logistics and Installation: Transporting and installing a door of this size and weight would be a major logistical undertaking, requiring heavy machinery and specialized expertise.

Are There Any Next Steps?

For an accurate quote, you would need to consult with firms specializing in large-scale, high-pressure blast or industrial doors, such as those that might supply doors for nuclear power plants or the military. One might like to contact specialized industrial door manufacturers like Krieger Specialty Products or Overly Door

Company. Both are specialized engineering teams that oversee blast-resistant or high-security projects. Next, you would need to provide them with specific operational requirements, the required pressure in PSI, and site details to get a detailed engineering assessment and cost estimate. The reality here is that none of us, less than billionaires, will ever be able to afford such a project, so there is not much value in wasting time on a pipe dream like this, unless there are a number of sites that are ready to go and costs can be distributed over many years and across many people. This distributed cost-sharing may make a sealed door of this magnitude more affordable for us all.

Chapter 10 connects spiritual, psychological, and practical approaches to survival during catastrophic, extinction-level events, suggesting that both spiritual preparation and physical planning are important. On the spiritual side, testimony such as Sister Grace Marie's near-death experience emphasizes faith, divine guidance, and the idea of protected refuges for believers. These safe places are described as divinely chosen locations where supernatural provision and protection are promised. This occurs especially during phases of societal collapse, persecution, or upheaval, all of which are likely to immediately precede and follow a solar nova/pole shift. Physically, the discussion shifts to the engineering and logistical realities of preparing for such events. Specifically, the construction of massive underground refuges with specialized, high-strength, watertight vault doors that could withstand extreme environmental threats like pole shifts, mega tsunamis, and ice ages. The engineering requirements highlight the prohibitive cost and technical complexity of building such structures, underscoring the challenges facing those seeking physical security. Both lines of thought converge on the notion that our survival, whether spiritual or physical, requires extraordinary commitment, resources, and a willingness to cooperate with others. The entire premise rests upon ultimately acknowledging that some aspects of survival may be beyond individual control, directing attention to the importance of being entirely prepared in body, mind, and spirit.

PART III
The Seers

Introduction

The ongoing observation of All Domain Anomalies (ADA), interpretations of pre-biblical and biblical encounters, the mystical experiences reported by both saints and modern seers, when examined in totality through a broader lens of comparative religious studies and transpersonal psychology, reveal patterns suggestive of a recurring divine triad.

It is this divine triad intervening in human affairs throughout history that we are to consider in Part III. All throughout church history, there have been what the Catholic Church usually calls saints. There is a lifetime of reading that one can do to learn more about them. These special people have often seen things or have heard messages from various divine sources, from St. Michael the Archangel, to Jesus, to Joseph, and the Virgin Mary, who are among the most popular of these encounters. Marian apparitions are among the most common and have captivated both believers and non-believers for centuries. There are untold numbers of stories of divine visitations and warnings reported on a global scale that have impacted many people over time, and not just Catholics. From Guadalupe to Lourdes and Fatima, similar events are woven into religious history, inspiring faith and fostering communities. What some modern researchers see is that there is a commonality between historical pre-biblical encounters, post-biblical encounters, the saints, modern Unidentified Anomalous Phenomena (UAP), and All Domain Anomalies (ADA). The patterns are suggestive of a recurring divine triad intervening in human affairs throughout history that has been secularized by governments for the purpose of control. The sciences have also collectively utilized the Galileo effect to argue that everything is pseudoscience and therefore not worthy of consideration.

Part III offers you a way to move your frame of reference, utilizing the striking parallels between mystic visions, certain UAP, and ADAs considering all the information found in Parts I and II of this book. In Part III, we will be examining the seemingly disparate phenomena of angels and ADA to demonstrate that they are more than connected; some of them are the same thing. Part III lays the foundation for how these differing ideas might share underlying characteristics or even represent diverse cultural interpretations of similar trans-dimensional or spiritual experiences. There is also a burgeoning thought on the warnings given, what they represent, and how one might incorporate their messages into a deeper faith today to grow spiritually into the higher being that God calls us to be. There are several writers who express these ideas with valid biblical proof.

Some writers use an academic hard examination of historical records at the Vatican. Other writers are expressing their own firsthand experiences with the divine. These authors attempt to show underlying connections through academic research, logic, and personal experience. The best way forward is to gain a greater understanding of how these connections will change your frame of reference by reading the works of the following authors. This is another jump-off point, and in no way is this an exhaustive or complete list, as there are other writers who have equally contributed to these ideas. This list is simply in alpha order by last name as follows:

1. Mauro Biglino
2. Chris Bledsoe
3. Randall Carlson
4. Edgar Cayce
5. Germaine Dieterlen
6. Ken Gooudsward

7. Steven Greer
8. Marcel Griaule
9. J. Allen Hynek
10. Carl Jung
11. Bob Lazar
12. George Knapp
13. Dianne Pasulka
14. Douglas Petrovich
15. Robert Schoch
16. Robert K.G. Temple
17. Jacques Vallée
18. Erich Von Däniken
19. Paul Wallace
20. Doug Vogt

These authors and theorists have found ways to overcome the Galileo Effect. The rest is up to you, dear reader, to become informed. You need to understand that we have all been lied to by many trusted organizations and governments. The information found in this book will open your eyes, and the opening of your heart is up to you to discover. Part III will attempt to bring together a unified process concerning all the previous sections concerning galactic cycles and solar cycles that impact us over eons of time. We will explore the connection of divine messages about the sun, and how the sun has, will, and always plays a significant role in sculpting our lives. Interpretations of Marian prophecies as global thermonuclear war are giving way to the idea that the messages reference the coming actions of the sun and reactions of the Earth. The seers are telling us that the processes involved in the various cycles that have changed the planet will continue to change the Earth via these same cyclical events forever into the future. Astronomical observations, scientific hypotheses, and the modern-day seers surrounding ADA events will be explored, raising the question, *"Could celestial warnings whispered through apparitions point towards a different kind of fire in the sky?"*

In Part III, we will examine the dialogue between faith, science, and spirituality by exploring how a re-interpretation of divine apparitions through the lens of a split in our timeline is required to foster a deeper understanding and appreciation for what is to come. Part III delves into the importance of open-mindedness, critical thinking, and a willingness to explore alternative explanations for phenomena that have historically been solely attributed to the supernatural. The role of hope and resilience in the face of potential global challenges will be explored, emphasizing the importance of finding meaning and purpose in both scientific discovery and spiritual belief. Part III will conclude by envisioning a future where the understanding of Marian apparitions as potential warnings of a split in our timeline can lead the way to a more unified and harmonious relationship between humanity, the cosmos, and the Creator. The potential for a deeper sense of stewardship for our planet and a renewed appreciation for the interconnectedness of all the things found in this book is what it is all about. Ultimately, Part III will offer a message of hope, suggesting that by embracing deeper faith, a greater understanding of science, and spirituality, humanity can navigate the challenges of the future and build a more resilient and spiritually enlightened society.

Chapter 11
Across History Before the Bible

To understand how these recurring patterns connect across time, it becomes necessary to look beyond surface interpretations and examine the deeper threads woven throughout human history. The sun has been the focus of many ancient civilizations. There are historians who believe that interactions with the divine have assisted humanity throughout time. The idea of All Domain Anomalies (ADA) is not limited to modern accounts, but instead appears globally across cultures, traditions, and recorded memory, extending far beyond contemporary seers and historical church figures. These encounters suggest an interconnected experience that spans civilizations, prompting a closer examination of culture and ancient texts, many of which predate the Bible and current scriptural canon. By exploring these early sources, a broader framework begins to emerge, revealing how the phenomena of ADA have been consistently interpreted and reinterpreted across time. The bottom line is that ADA is an unknown technology that is flying machines.

There are several accounts of flying machines or craft, often depicted as magnificent chariots or palaces used by deities, heroes, and even mortals.

Ancient Names Associated with Flying Craft:

Mesopotamia

In Sumerian culture and mythology, the *"Shining Ones"* are primarily associated with the Anunnaki.

Divine beings: The Anunnaki were a group of deities revered by the ancient Sumerians, Akkadians, Assyrians, and Babylonians.

Origin and meaning: The name *"Anunnaki"* are derived from An, the Sumerian god of the sky, and can be translated as *"princely offspring"* or *"offspring of An"*. They were believed to be the children of An, who is the god of the heavens, and Ki, who is the goddess of Earth.

Associated with Wisdom and Power: The *"Shining Ones"* were associated with wisdom, power, and enlightenment, often linked to celestial bodies like the sun and stars.

Role in Mythology: In Sumerian mythology, the Anunnaki held the vital role in decreeing the fates of humanity.

Bringers of Knowledge: They were also believed to have brought advanced knowledge and civilization to humanity.

Connection to Other Beliefs: Some theories attempt to connect Anunnaki with ideas of ancient aliens who may have influenced early civilizations. In Sumerian mythology, the term *"sky people"* refers to the Anunnaki.

According to Sumerian beliefs, the *"sky people,"* or *"shining ones,"* were initially associated with the heavens, but later texts suggest an increased role in earthly matters. Over time, particularly during the Old Babylonian Period, the name Anunnaki was specifically applied to the deities of the underworld. The Igigi were associated with the heavens. In some texts, they are described as humanoid beings who came to Earth, possessed vast knowledge, and lived for hundreds of thousands of years. Some more modern researchers interpret the Anunnaki as extraterrestrial beings who visited Earth in ancient times.

They believe the Anunnaki played a role in the creation of humankind, potentially through genetic manipulation, and taught them various skills like agriculture and writing. These theories are largely based on the interpretations of cuneiform texts. The Semitic peoples are descendants of these areas who had knowledge of these stories. There are some who see parts of the Old Testament as a paraphrasing or re-telling of these older stories when words like

Elohim are properly translated. To learn more about Abraham, and how his stories in the bible relate to Sumerian mythologies, then find the works of Paul Wallace and Mauro Biglino. Between the two of them, you will find an entire world of thought-provoking ideas to consider.

India

Vimanas: These are flying chariots or palaces mentioned in ancient Indian texts like the Vedas and epics like the Ramayana and Mahabharata. They are described in detail, including their use in warfare, ability to fly within the Earth's atmosphere, and even travel into space and underwater. Some famous examples include:

- *Pushpaka Vimana:* A celestial self-propelled car in the Ramayana, capable of traveling anywhere at will. It was originally crafted by Vishwakarma for Brahma and then given to Kubera before being stolen by Ravana.
- *Saubha:* A flying machine owned by the ancient Indian king Salva, described as being able to become invisible and maneuver at will.

Various Other Cultures: Many cultures have myths of gods and heroes riding in flying chariots as follows:

- *Indra's Chariot (India):* Used in battles and for celestial travel, according to the Mahabharata.
- *Helios' Sun Chariot (Greek Mythology):* Pulled by winged dragons or horses, it carries the sun god across the sky each day.
- *Thor's Goat-Drawn Chariot (Norse Mythology):* A flying chariot pulled by goats in Norse legends.
- *Hebo's Chariot (Chinese Mythology):* A chariot pulled by two dragons.

- *Magic Carpets (Middle Eastern Folklore):* Known from tales like *"One Thousand and One Nights"* and associated with figures like King Solomon. King Solomon reportedly possessed a large, fast-flying carpet that could carry many men. Some believe these stories have roots in Jewish folklore.
- *Alexander the Great's Flying Machine (Legend):* In a legend, Alexander the Great fashioned a device powered by griffins and meat, allowing him to ascend to the heavens.
- *Daedalus' Wings (Greek Mythology):* While not a craft in the sense of a machine, Daedalus created wings for himself and his son Icarus by fastening feathers together with thread and wax, enabling them to fly.

Based on existing archaeological findings and interpretations of indigenous legends and oral traditions from North, Meso, and South America, it is somewhat more challenging to provide a definitive list of names for *"flying craft"*. The ancient American cultures are understood in modern terms, such as airplanes or spaceships. However, some ancient artifacts and mythological figures can be interpreted as representations or descriptions of flying objects or beings with flight-related attributes.

Mesoamerica

- *Quetzalcoatl:* In Aztec culture, a feathered serpent deity and creator god, sometimes depicted with anthropomorphic forms such as Ehecatl, the wind god.
- *The Voladores (Pole Flyers):* An ancient Mesoamerican ceremony, likely originating with the Nahua, Huastec, and Otomi, involving participants descending from a tall pole. The participants would descend from the pole while tied with ropes. This ritual suggests a focus on the concept of flight, although not with a mechanical craft.

- *Olmec Monument 19 "Rocket Man":* A controversial interpretation of an Olmec carving depicting a figure within a serpent-like enclosure, which some interpret as a "rocket man" or a figure within a flying craft.

South America

- *Quimbaya artifacts/Tolima fighter jets:* A collection of gold artifacts found in Colombia that are associated with the Quimbaya culture, from 300 to 1550 CE. Some have interpreted it as ancient portrayals of aircraft.
- *Inca artifacts:* Some Inca gold artifacts, like the Quimbaya artifacts, have been interpreted by some as representing ancient flying machines. The controlling narrative here is that these represent stylized animals. Could it be that the stylized animals are a cover so that mainstream archaeology does not have to deal with challenging questions of its narrative and control?

North America

- *Flying Head (Kanontsistonties):* A creature from Iroquois folklore, described as a giant head with long hair, terrifying eyes, and sharp fangs, sometimes depicted with wings and bird legs. Some interpretations suggest a connection to ideas of flight or beings capable of aerial movement.
- *Thunderbird:* A powerful mythical bird in various North American Indigenous cultures, often associated with storms and other supernatural powers.
- *White Buffalo:* In several Native American religions, particularly those of the Lakota and other Plains Indians, the white buffalo, or bison, is considered sacred and spiritually significant. Its birth is viewed as a prophetic sign, signifying hope, abundance, and good times to come. This belief stems from the legend of the White Buffalo Calf Woman, who

brought a sacred pipe and taught the Lakota seven sacred ways to pray. The significant connection here is that for Catholics, there are seven sacraments, and this is no mere coincidence. The story says she promised to return as a white buffalo calf to restore harmony and spirituality to the world. Some tribes also see the white buffalo as a warning, reflecting the vulnerability of Mother Earth and the need to achieve balance and respect for all living things. The White Buffalo holds deep spiritual meaning as a sacred omen of hope and change.

- *Kachinas (or Katsinas):* Kachinas are spiritual beings in the Hopi culture and are revered as sacred spirits among the Hopi people.

They are believed to influence the natural world, bringing blessings like rain and healthy crops, and serving as messengers between humans and the divine. Kachinas are spiritual beings in Hopi cosmology, represented by masked dancers and carved dolls that serve as tools for education and connection to the spiritual realm. Kachina dolls, carved from cottonwood roots, are representations of these spiritual beings. These dolls are traditionally given as gifts to young girls to educate them about the different Kachinas, their characteristics, powers, and the stories and beliefs attached to them. Kachina impersonators are men initiated into various clans who represent these spirits. Kachina impersonators use masks and dances in the plazas of Hopi villages during the Kachina season. Each Kachina doll is intricately carved and painted, embodying specific characteristics, powers, and teachings of the spirit it represents.

Africa

The Dogon people of Mali, West Africa, have a complex and fascinating tradition surrounding their connection to the Sirius star system. At the heart of this tradition is the belief that their ancestors were visited by extraterrestrial beings called the Nommo, who are said to have originated from the Sirius system.

Key Traditions and Beliefs Related to the Dogon:

- *The Nommo and their arrival:* The Dogon believe that the Nommo, described as amphibious, fish-like beings, descended from the sky in a vessel or *"ark"* accompanied by fire and thunder. They are referred to as *"Masters of the Water," "Monitors,"* and *"Teachers."* These Nommo are said to have brought advanced knowledge to the Dogon ancestors, including information about the Sirius star system, agriculture, weaving, and cosmology. This knowledge is the understanding of Sirius B, the dense white dwarf companion star to Sirius A, which is invisible to the naked eye and was only observed by Western astronomers centuries later.
- *Astronomical knowledge and beliefs:* The Dogon refer to Sirius A as *"Sigi Tolo"* or the Star of Sigui, and its companion as *"Po Tolo"* or Digitaria star. They describe Sirius B as orbiting Sirius A in an elliptical path over a 50-year period, which is scientifically accurate.

They also claim knowledge of a third star in the Sirius system, Sirius C, which remains unconfirmed by modern astronomy. The Dogon cosmology also includes the knowledge that Earth and other planets revolve around the sun and rotate on their respective axis. They are said to know about Jupiter's four main moons and Saturn's rings, celestial features typically observable only with telescopes.

Transmission of Knowledge and Rituals:

The Dogon have preserved this advanced knowledge through oral tradition passed down through generations. Their beliefs are celebrated and transmitted through ceremonies, symbols, art, and carvings. The Sigui ceremony, held every 60 years that corresponds to the orbital period of Sirius B, is a significant ritual associated with passing on this knowledge. During this long and complex ceremony, which can last for several years, younger men isolate themselves and speak in a secret language as they prepare to receive this knowledge from their elders.

The "Sirius Mystery" of the Dogon: French anthropologists Marcel Griaule and Germaine Dieterlen documented the Dogon's knowledge in the 1930s and 1940s. Griaule and Dieterlen revealed the unexpected depth of Dogon knowledge, and that then sparked the publishing of the book "Sirius Mystery." Robert K.G. Temple popularized the idea of extraterrestrial contact with his 1976 book, *"The Sirius Mystery"*. Skeptics propose alternative explanations, such as cultural contamination through contact with Europeans or the inheritance of knowledge from ancient Egyptian astronomical traditions. The Dogon language, for example, is said to contain ancient Egyptian words relating to the Sirius system. Despite ongoing debates and a lack of definitive evidence, Dogon's traditions concerning their ancestors' encounters with beings from Sirius highlight a remarkable understanding of the cosmos within their cultural framework. These examples highlight the universal human fascination with the divine and with flight. They often attribute the ability to fly to deities or powerful individuals using magical or advanced devices beyond human comprehension at that time in history.

Chapter 12
Mesopotamian/Semitic Peoples/Göbekli Tepe

Understanding when key events occurred in the ancient world provides essential context for interpreting archaeological sites and historical developments with greater clarity that can then be applied to the current evidence for solar nova. While multiple locations contribute to this evolving timeline, no sole source captures the full scope of humanity's early past, making it important to view these findings as part of a broader, incomplete picture. The purpose of this timeline is not to present a definitive record but to highlight the depth and complexity of our ancient origins, allowing connections between biblical periods and earlier civilizations to become more apparent. By examining these chronological markers, the reader gains a clearer perspective on how these civilizations developed and how their influence continues to shape modern understanding. There are several sources to do your own research, as follows:

- Encyclopedia.com
- Humanities LibreTexts
- PoliLingua Translation Agency
- Study.com
- The Metropolitan Museum of Art
- The OER Project

The following timeline highlights key periods and developments related to Mesopotamia, Semitic peoples, and the ancient site of Göbekli Tepe. It is important to note that Göbekli Tepe predates the Mesopotamian civilizations but offers insights into the wider ancient Near East.

Pre-Neolithic: From 11,000 BCE to 8,000 BCE: Göbekli Tepe flourishes in southeastern Anatolia or modern-day Turkey, highlighting monumental stone structures built by young civilizations, challenging conventional understanding of early human capabilities. The site, featuring T-shaped pillars with carved animal reliefs, possibly served as a ritual or pilgrimage site.

Neolithic & Early Bronze Age: From 10,000 BCE: Marks the beginnings of agriculture in the Middle East.

7000 BCE - 5500 BCE: Pottery Neolithic period in Mesopotamia, with early pottery styles like Hassuna, Halaf, and Samarra.

5500 BCE - 4000 BCE: Ubaid period, characterized by distinctive painted pottery and the emergence of temple architecture.

4100 BCE - 2900 BCE: The Uruk period sees the rise of cities, monumental temples, the development of cuneiform writing, and cylinder seals.

3500 BCE: Semitic languages are believed to have emerged in the ancient Near East.

2900 BCE - 2350 BCE: Early Dynastic period in Sumer, marked by rival city-states like Uruk, Ur, and Lagash.

2334 BCE - 2154 BCE: Akkadian Empire, the first empire to unite Mesopotamia, established by Sargon of Akkad. Akkadian, a Semitic language, becomes dominant.

Middle & Late Bronze Age: From 2112 BCE - 2004 BCE, the third Dynasty of Ur (Ur III), sees a resurgence of Sumerian culture and the building of ziggurats.

2025 BCE - 609 BCE: Assyrian Empire, initially centered on the city of Ashur. Assyrians are Akkadian-speaking Semitic people.

1792 BCE - 1595 BCE: Old Babylonian Empire, with its capital at Babylon, reaches its height under Hammurabi, known for his law code.

1600 BCE - 1160 BCE: The Kassite period sees the Kassites dominating Babylonia.

1400 BCE - 1000 BCE: Middle Assyrian period.

1100 BCE Onwards: Nomadic peoples like Aramaeans and Chaldeans enter Mesopotamia, impacting existing kingdoms. Aramaic is a Northwest Semitic language.

Iron Age & Late Periods, From 900 BCE - 612 BCE: The Neo-Assyrian Empire expands and asserts control over a vast territory, including Mesopotamia and parts of the Near East.

626 BCE - 539 BCE: Neo-Babylonian Empire emerges, famously conquering Jerusalem and exiling the Israelites.

539 BCE: Fall of Babylon to the Persian Achaemenid Empire under Cyrus the Great, marking the end of independent Mesopotamian rule.

332 BCE: Alexander the Great conquers Mesopotamia, beginning the Hellenistic period.

Post-Classical 75 CE: The latest known Akkadian documents are dated.

Modern Day: Semitic languages like Arabic, Hebrew, Amharic, and Aramaic continue to be spoken, highlighting their enduring legacy.

In Chapter 12, we have explored a brief history of time to set the stage for ancient myths and texts from various cultures. We have demonstrated that many ancient texts, predating the Bible, contain accounts of flying machines or crafts used by deities and heroes. We have considered examples of these connections and conclude that these examples show a universal human fascination with flight and the divine when considered with the information in Chapter 11. Many cultures have attributed the ability to fly to deities or powerful individuals using devices beyond human comprehension at the time they are witnessed. Finally, Chapter 12 concludes with a timeline of Mesopotamian and Semitic peoples, noting how the ancient site of Göbekli Tepe predates these civilizations. The connection to the sun will become clearer over time as you go deeper into this book. Establishing a clearer perspective on how these civilizations developed and how their influence continues to shape modern understanding will help the reader to see the connections that lead us in a great circle back to the sun.

Chapter 13
Ancient Trinitarian Considerations

Building on the connections established in the previous chapter between humanity's fascination with flight, ancient timelines, and the divine, this chapter shifts focus toward the structure of religious belief systems, particularly the role of women within ancient Egyptian spirituality. This is particularly important because it will bring us to the divine feminine. In this context, women were not merely participants but held significant positions that reflected deeper theological principles, including the presence of divine triads within their cosmology. Although religious institutions were often male-dominated, women served as priestesses, ritual participants, and spiritual intermediaries, highlighting a more complex and balanced understanding of power, divinity, and sacred roles within ancient society.

Pre-biblical Trinitarian Ideas:

Goddesses and the Divine Feminine: Ancient Egyptian religion featured a rich pantheon of goddesses who played crucial roles in creation, protection, and the afterlife.

Goddesses like Isis, Hathor, and Neith were powerful deities, embodying motherhood, love, wisdom, and justice. The presence of powerful goddesses underscored the importance of the feminine principle in their cosmology and religious beliefs.

Women as Priestesses in Religious Roles: While the priesthood was generally male-dominated, women served as priestesses, especially in the cults of goddesses. Their duties included performing rituals, making offerings, maintaining temples, caring for statues, and assisting worshippers. Priestesses played a dominant role in the priesthood of Hathor.

God's Wife of Amun: This powerful title, held by royal women, often signified a high-ranking priestess who was strongly associated with the god Amun. This role was most popular during the New Kingdom and Third Intermediate Period. The New Kingdom was ancient Egypt's *"golden age"* of empire and wealth, followed by the Third Intermediate Period, an era of political fragmentation and foreign rule. The God's Wife of Amun was responsible for performing rituals and offerings to Amun, and sometimes wielded significant political influence.

Professional Mourners: Women were employed as professional mourners, expressing grief at funerals through ritualized lamentations, a role considered specific to women to help the deceased transition to the afterlife.

Musicians and Dancers: Women played vital roles in temple rituals as singers, dancers, and musicians, providing entertainment for the deities. Women served as musicians, dancers, and chantresses or temple singers, particularly for goddesses like Hathor. Their performances were considered essential for appeasing the deities and maintaining cosmic harmony.

Dream Interpreters and Oracles: Women were also associated with dream interpretation and were consulted for guidance and insight into the future. Women were often more connected to intuitive thoughts than they were valued for.

Balance and Equality: The concept of balance and harmony was central to Egyptian religious thought, and this was reflected in the roles of men and women in religious life.

The position of the God's Wife of Amun was often balanced by the High Priest of Amun. These positions demonstrated an attempt to maintain equilibrium between the masculine and feminine aspects of the divine. While men held more public and administrative roles in religious institutions, women's contributions were equally valued and respected.

Beyond the Temple: Women also participated in religious life through personal rituals, offerings, and participation in festivals. The religious experiences of ordinary women, both in public and private life, were diverse and significant. The ancient Egyptian understanding of women in religion reflects a society that, while patriarchal in some respects, also recognized and revered the divine feminine principle.

Household Religious Practices: It is believed that women were responsible for maintaining personal shrines within their homes and offering prayers and thanks to the gods there. However, there were some limitations on women's roles.

- *Administrative positions:* While priestesses conducted rituals, they generally did not hold administrative positions within the temple hierarchy, unlike male priests.
- *Access to certain spaces:* Some evidence suggests that women may have had limited access to the inner sanctuaries of temples, particularly during menstruation.

Overall, ancient Egyptian society viewed the roles of women in religious practices as important for maintaining balance and harmony within the cosmos and society itself. Their involvement spanned various roles, from priestesses and influential figures to musicians, mourners, and household caretakers of religious life. Ancient Egyptians did not have a single, fixed concept of a divine trinity like the Christian one, but they frequently used triads of deities to represent a complete system or a family dynamic. These triads often symbolized concepts like creation, death, and rebirth, or aspects of a single deity. While not identical to the Christian Trinity, the Egyptian use of triads influenced later religious thought, particularly the concept of a divine family or a unified representation of divine power.

Triads as A Symbolic Unit: In ancient Egyptian religion, the concept of a divine family or *"trinity"* was commonly expressed through groups of three gods known as triads. Ancient Egyptian religion often grouped deities into triads. These triads were not necessarily seen as three distinct gods but rather as a unified representation of a concept or a divine family. These triads typically consisted of a father, a mother, and their son, and were worshipped at major cult centers throughout Egypt. The ancient Egyptian concept of a triad is not identical to the Christian idea of the Trinity, but there are some intriguing similarities that need consideration.

- *Family Structure:* Egyptian triads were often a literal family unit consisting of a father, mother, and child, representing a fundamental family structure and its importance within society.
- *Symbolism of Three:* The number three itself held significance in ancient Egyptian thought, representing plurality and completion. Triads were seen to group and understand the multifaceted nature of the gods.
- *Aspects of a Single God:* In some cases, the male gods within a triad could represent distinct aspects or stages of a single god's existence, emphasizing the eternal cycle of death and rebirth.
- *Difference from Christian Trinity:* While both concepts involve three divine entities, the Egyptian triads were not typically understood as three persons in one God, as is the case in the Christian Trinity.

Important Triads in the Ancient World: The roles of the Egyptian goddesses within these triads, particularly their maternal role, are not unlike the role of the Holy Spirit in the Christian understanding of the Trinity.

Osiris, Isis, and Horus: This triad, centered at Abydos, is perhaps the most famous, illustrating the themes of resurrection, magic, and royal succession. This triad is one of the most well-known, with

Osiris as the king, Isis as his wife and sister, and Horus as their son. Osiris is often associated with death and resurrection, Isis with motherhood and magic, and Horus with kingship and the sky.

Amun, Mut, and Khonsu: In Thebes, the triad consisted of Amun, the hidden god, Mut, his wife, and Khonsu, their son. This triad represented a family dynamic and the interconnectedness of distinct aspects of the divine. This triad represented distinct aspects of the same divine power.

Ptah, Sekhmet, and Nefertem: Worshipped at Memphis, this triad also highlighted the creator god Ptah, his warrior goddess consort Sekhmet, and their son Nefertem. Some other examples include the triad of Atum, Shu, and Tefnut, representing creation and the first divine couple. Ancient Egyptian deities were not fixed or static. They could be combined or syncretized with other deities, and their roles and relationships could shift over time. This fluidity allowed for multiple interpretations and representations of divine triads.

Chapter 14
The Forgotten Maternal Nature of the Divine

Hathor emerges as one of the most complex and multifaceted figures in ancient Egyptian religious practice, embodying themes of motherhood, creation, and cosmic order. Although she is not explicitly documented prior to her emergence in historical records, evidence suggests that her origins are deeply rooted in earlier predynastic beliefs and related deities that gradually evolved into her recognized form. Her earliest clear depictions appear during the Old Kingdom, yet her attributes indicate a synthesis of earlier symbolic traditions connected to the sky, fertility, and the divine feminine. This progression reflects the fluid nature of ancient belief systems, where deities developed over time through cultural integration and reinterpretation. Chapter 14 is a closer look at these origins and how the patriarchy of Israel left the maternal nature of the divine behind.

<u>*Pre-Egyptian Origins of Hathor*</u>:

Early Cow Goddesses: Hathor's association with cows suggests a connection to earlier Predynastic cow goddesses like Bat and Mehet-Weret, who were also linked to the sky and motherhood.

Sky Goddesses: Hathor's association with the sky, particularly the Milky Way, connects her to earlier sky goddesses like Nut, who was also depicted as a cow.

Absorbed Attributes: Hathor's worship likely absorbed elements from these earlier deities, evolving into the complex goddess known in historical Egyptian times.

Early Depictions: While clear depictions of Hathor appear later, some artifacts from the Early Dynastic Period between 3100 and 2686 BCE may refer to her.

Name Origin: The name *"Hathor"* itself, meaning *"estate of Horus,"* may not be her original name, further suggesting a gradual development and assimilation of earlier concepts. The origins of Hathor are somewhat fluid and debated among Egyptologists.

In Egyptian mythology, Hathor was a significant goddess with a multitude of roles, including goddess of the sky, women, fertility, love, beauty, music, dance, and joy. She was often depicted as a cow, symbolizing her maternal and celestial nature, or as a woman with cow ears or horns and a sun disk between them. While Hathor is definitively documented from the Old Kingdom period onwards, there's evidence suggesting her cult may have existed even before this time. Pre-dynastic Egyptian artwork frequently features cattle and women with upraised, curved arms that resemble horns. These upraised arms or horns could point to the veneration of cattle goddesses like Hathor. One specific example is the Gerzeh Palette from around 3500-3200 BCE.

Gerzeh Palette: A stone palette from the Naqada II period between 3500-3200 BCE that features a cow's head with inward-curving horns surrounded by stars. It was used for grinding cosmetics as indicated by traces of malachite found on its reverse side. It is thought to depict or be associated with the cow goddess, potentially Bat or Hathor. The stars on the palette connect the bovine figure with the sky, suggesting an early form of Hathor worship. However, the horns on these early representations are typically depicted as curving inward, whereas Hathor's horns are usually shown curving outward. The Palette is made of schist, has an oval shape that widens slightly at the top, and is pierced so that it can be hung. One side features a stylized relief of a cow's head with upward-curving horns and prominent ears. Upon the Palette are five stars depicted with two at the tips of the horns, one on the top of the head, and two on either side of the ears. This imagery likely evokes one of the bovine-form deities, most likely Hathor, who identified with the heavens during this period.

The five stars represented the souls of the deceased in the afterlife who are seeking everlasting life. Hathor eventually supplanted earlier deities, including a crocodile god at Dendera and the goddess Bat, whose cult she absorbed. Hathor's association with the sun god Ra also solidified during the Old Kingdom as he became increasingly central to Egyptian royal theology.

Possible Association with Bat: The Narmer Palette features a bovine deity with inward-curving horns. This palette is a significant ancient Egyptian ceremonial stone tablet, dating to around 3100 BCE, which features some of the earliest hieroglyphic inscriptions ever discovered. The Narmer Palette is most famous for its intricate carvings that are widely believed to depict the unification of Upper and Lower Egypt under King Narmer. It also shows a very ancient cow goddess, possibly the earliest of the sky cow deities. Some scholars believe this may represent Bat, an earlier cow goddess later absorbed by Hathor.

Bat is an ancient Egyptian cow goddess associated with Upper Egypt. She was depicted as a human face with cow ears and horns or as a woman. There is evidence of early worship of Bat from the earliest records of ancient Egyptian religious practices. It is believed that these practices possibly originated in the late Paleolithic cattle herding period. The name Bat may be linked to the word *"ba,"* often translated as "soul" or *"personality"*. Bat was later overtaken by Hathor during the Middle Kingdom, when her identity and attributes merged with those of the goddess Hathor, a similar goddess worshipped in a nearby region. The rise of Hathor included the imagery of Bat that persisted throughout ancient Egyptian history, particularly on the sistrum, a sacred musical instrument. The bat was associated with the Milky Way, seen as a pool of cow's milk that illuminated the way to eternal life. She possessed powerful magic, allowing her to perceive both the past and the future.

Mehet-Weret: A celestial cow goddess representing the *"great flood"* or the primordial waters from which the sun god re-emerged.

She is also associated with rebirth and the sky, often depicted with the sun disk between her horns. Mehet-Weret is an ancient Egyptian deity of the sky, associated with streaming water, creation, and rebirth. The name Mehet-Weret means *"Great Flood"* or *"Great Tide"*, which refers to her connection to the primeval waters.

- *Role in Creation:* In ancient Egyptian creation myths, she gives birth to the sun at the beginning of time. She is portrayed as a cow with a sun disk between her horns. She is associated with goddesses like Neith, Hathor, and Isis, all sharing similar characteristics. She is also identified with the Wedjat, or the Eye of Horus or Ra.
- *Absorbed by Hathor:* She is thought to have lacked an independent cult, and her sky attributes were likely absorbed by Hathor as early as the Old Kingdom.
- *Milky Way Connection:* She was also probably associated with the Milky Way in the night sky.

Nut: Another cow goddess, often depicted as a cow whose body forms the sky. She is the mother of the sun god Re and the gods of the sky. Nut is pronounced *"newt"*. Nut is the ancient Egyptian goddess of the sky and heavens. She was usually depicted as a woman bent over the Earth, who is her brother, and husband Geb, with her head in the West and feet in the East. She was also sometimes shown as a celestial cow. She played a crucial role in the daily cycle of the sun, swallowing it as it set in the west and giving birth to it again at dawn.

- *Role in cosmology:* In Egyptian cosmology, Nut protected the Earth from being inundated by the surrounding waters.
- *Milky Way Connection:* Her depiction as an arched woman, sometimes studded with stars, evokes the Milky Way arching across the sky.

- *Children of Nut:* On five special days preceding the New Year, Nut gave birth to the deities Osiris, Horus, Seth, Isis, and Nephthys. These gods, except for Horus, were commonly referred to as the *"children of Nut."*

Foreign Influence (Baalat): Some scholars suggest Hathor may have originated as a Semitic goddess, possibly Baalat, who was worshipped in Canaan. One theory proposes that Hathor might have originated as the Canaanite goddess Baalat, and then became integrated into Egyptian religion during the second millennium BCE. In this scenario, Hathor would have been considered an Egyptianized form of a Semitic goddess.

An inscription found in the Egyptian mining area at Serabit el-Khadem in the Sinai mentions Baalat. There is evidence of trade between Egyptian miners and the Semitic peoples inhabiting the region. This suggests the possibility of cultural exchange where the Egyptians might have adopted aspects of foreign goddesses into their own religious system. Regardless of her exact origins, Hathor's prominence increased significantly during the Fourth Dynasty of the Old Kingdom. While Hathor is not clearly identified before the Old Kingdom, predynastic Egyptian artwork and neighboring cultures show veneration of similar deities. It is possible that Hathor developed by incorporating aspects of these earlier goddesses or through cultural exchange. Bat, Mehet-Weret, and Nut are ancient Egyptian cow goddesses associated with the sky and creation. The Gerzeh Palette is a Predynastic cosmetic palette featuring a bovine figure, potentially representing Bat or Hathor. The Narmer Palette is a significant ancient Egyptian ceremonial stone tablet, which features some of the earliest hieroglyphic inscriptions ever discovered.

Trinitarianism and the Divine Feminine:

The concepts of Trinitarianism and the Divine Feminine, though stemming from distinct religious and theological traditions, offer interesting points of comparison and potential correlations worth exploring. Trinitarianism, as understood in Christian theology, centers on the belief in one God existing as three coequal persons of the Father, Son, and Holy Spirit. The Divine Feminine encompasses the feminine aspects of the divine or God that appear in various traditions. The Divine Feminine is often symbolized by goddesses or feminine principles like nurturing, intuition, and creation. Examining their similarities and correlations can enrich our understanding of the diverse ways humans perceive and engage with the sacred.

Similarities in Attributes: One key area of comparison lies in the shared attributes attributed to both Trinitarian persons and the Divine Feminine. For instance, God the Father, in the Trinitarian understanding, is often associated with the ultimate source of all being and creation. Similarly, the Divine Feminine in various traditions is revered as a generative force.

The divine force is responsible for bringing forth life and sustenance, often symbolized by Mother Earth or other life-giving aspects of nature. The Holy Spirit within Trinitarian theology is viewed as the life-giving, inspiring, and indwelling presence of God, who dwells within believers as the temple of the Holy Spirit. Some scholars have noted the use of feminine imagery in some biblical passages to refer to the Holy Spirit. One example of feminine imagery is the description of God as a nursing mother in Isaiah 66:13. This resonates with certain aspects of the Divine Feminine, which often emphasizes nurturing, healing, and intuitive wisdom. Early Christian writers also equated the Holy Spirit with Sophia, which is a feminine name for Wisdom that is grammatically feminine in Greek.

Correlations in Roles and Functions: Another area to consider involves the correlations between the roles and functions within Trinitarian theology and the expressions of the Divine Feminine. The concept of perichoresis, for instance, emphasizes the mutual indwelling and interdependence of the three persons of the Trinity. Perichoresis highlights the dynamic relationality at the heart of the divine life.

This relationality can be seen as mirroring the emphasis on community, connection, and collaboration often associated with the Divine. Additionally, the role of Mary as the mother of Jesus in Christianity, revered as the *"Mother of God"* or *"Queen of Heaven"* in some traditions, can be seen as a point of connection with the reverence for feminine figures in the Divine Feminine, even if her role within Trinitarian theology is distinctly not part of the Trinity itself. The tenderness and dedication attributed to Mary resonate with the nurturing and compassionate qualities often associated with the Divine Feminine.

Potential for Mutual Enrichment: While acknowledging the distinct nature of Trinitarian theology and the diverse expressions of the Divine Feminine, exploring their similarities and correlations can open avenues for richer theological reflection and spiritual understanding. Recognizing the presence of feminine imagery and attributes within Trinitarian concepts can challenge potentially narrow or exclusively masculine interpretations of the divine, fostering a more balanced understanding of God.

Likewise, a deeper appreciation of the Divine Feminine can inform and deepen our understanding of the nurturing, life-giving, and relational aspects within the Triune God. Ultimately, this exploration can contribute to a more holistic and encompassing appreciation of the divine, embracing the fullness of its mystery in ways that resonate with our human experiences and spiritual journeys.

Judaism, a Frame of Reference:

Throughout history, religious traditions have been shaped not only by belief but also by the cultural structures surrounding them. Judaism has often been viewed as a patriarchal system, influenced by its foundational texts and the societal frameworks that developed alongside them.

Biblical Narratives and Law: The Torah and other biblical texts frequently feature men as the central figures, protagonists, and lawmakers. These texts often gave men the prerogative in matters of marriage and divorce, while women were largely dependent on their fathers and husbands.

Traditional Gender Roles and Exclusions: In traditional Judaism, men were primarily responsible for public religious activities like communal study and prayer, while women's roles were more often centered in the home. Women were traditionally excluded from participating in public religious rituals in the same way as men. One example is that women were not counted as part of a minyan, or prayer quorum, nor were they called to read from the Torah in some settings.

Societal Norms and Interpretation: The cultural contexts in which Judaism developed, particularly in its initial stages, were often patriarchal, shaping the interpretation and application of religious laws and traditions. Some traditional interpretations of Jewish texts have emphasized hierarchical distinctions between men and women, associating men with the public sphere and women with the domestic realm. This separation has developed over time into cultural subservience that extended all the way up to God as solely a masculine archetype.

Varied Perspectives within Judaism: Different Jewish movements and communities have diverse perspectives on gender roles and equality.

Reform and Reconstructionist Judaism, for example, have been at the forefront of advocating for and implementing changes that promote gender equality.

Historical Evidence of Women's Influence: Despite patriarchal structures, Jewish women have had significant roles in shaping Jewish life throughout history. For instance, women's movements from around the mid-20th century spurred changes in many Jewish denominations, leading to greater equality in areas like rabbinic ordination and communal leadership.

In Chapter 14, we considered the significant and diverse roles of women in ancient Egyptian religious practices, challenging the idea that women in ancient societies were always subordinate. Women served as priestesses, particularly in the cults of goddesses like Hathor, as noted in Chapter 14, and held powerful titles such as *"God's Wife of Amun"*. They were also involved in rituals as musicians, dancers, and professional mourners, and were associated with dream interpretation and oracles. While men held more public and administrative positions, women's contributions were valued for maintaining cosmic balance and harmony.

Chapter 14 also explores the ancient Egyptian use of divine triads, which were groups of three deities, often a father, mother, and son, which symbolized a family dynamic or a unified representation of a concept of divinity. Famous examples include the triads of Osiris, Isis, and Horus; and Amun, Mut, and Khonsu. These triads are noted for being different from the Christian concept of the Trinity, as they were not three persons in one God. Chapter 14 then delves into the origins of the goddess Hathor, suggesting she evolved from earlier Predynastic cow goddesses like Bat and Mehet-Weret and absorbed their attributes. Early artifacts like the Gerzeh Palette and the Narmer Palette may depict these earlier bovine deities.

Chapter 14 then, goes on to compare and contrast the Christian concept of Trinitarianism with the Divine Feminine. While distinct, the two do share several similar attributes and correlations. One such similarity is that the Holy Spirit's feminine imagery and nurturing aspects both resonate with a Divine Feminine emphasis on creation and life-giving forces.

Finally, the chapter touches on the patriarchal structures in traditional Judaism, contrasting them with the diverse perspectives and movements for gender equality that have emerged over time, and how, from a historical perspective, the equality of roles culturally has moved away from equal to a masculine-controlled narrative.

Chapter 15
A New Frame of Reference

As the ideas developed in the previous chapters begin to come together, they invite a shift in perspective on the modern understanding of the Trinity. What is commonly understood as God in the form of Father, Son, and Holy Spirit, known through faith and revelation, appears distinct from earlier concepts such as the divine masculine, divine feminine, or sacred family structures found in ancient belief systems and some modern interpretations. These differences may not have always existed in such separation but rather developed over time through historical and cultural influences, including patriarchal frameworks that shaped how these ideas were interpreted. This raises an important question: are these differing views truly separate, or are they expressions of the same underlying supernatural concept viewed through different lenses across history?

Two Sides of the Same Coin:

Gender and Transcendence: For the faithful, God is a being of pure spirit that transcends human concepts of sex and gender. While the father is referred to in masculine terms and his love can be depicted with maternal images, the divine nature itself is beyond gender.

The Trinity: The Trinity is the central doctrine of faith, signifying the truth that there are Three Persons within the unity of God: Father, Son, and Holy Spirit. These are truly distinct Persons who are coequal and co-eternal.

No Feminine Principle in the Trinity: The official Catholic stance is that there is no feminine principle in the Trinity itself.

Some early Church thinkers did explore feminine attributes or metaphors associated with the Holy Spirit, but this was not adopted as official doctrine.

The Holy Family: The Holy Family of Joseph, Mary, and Jesus is an important model of family life and holiness in Catholicism.

However, it is a human family, not a representation of the divine nature of the Trinity in a traditional sense. For Catholics, the Holy Family is quite different from the ancient sense that came before it. Could it be that this is a form of control instituted by those in power to control the narrative and demand obedience?

Divine Masculine - Divine Feminine: The concept of *"Divine Masculine"* and *"Divine Feminine"* often refers to archetypal energies or principles that may or may not be personified, and which can be found in various spiritual traditions outside of and prior to Catholicism.

Catholic Understanding of God's Relationship with Humanity: God is called *"Father"* as the source of creation and transcendent authority, as well as for his goodness and loving care.

God's parental tenderness can also be expressed through the image of motherhood, emphasizing his immanence and intimacy with creation. While both Hebrew and Christian patriarchy use masculine imagery to show His separateness from creation, He is limited by the human ideology that the male gender is superior and not coequal. And so Why? Because the Catholic Church, and Christianity in general, does not adhere to a doctrine of a *"divine feminine"* in the same way that some other spiritual traditions or ancient belief systems do. The Catholic understanding of the Trinity is primarily male, a divine mystery that is distinct from concepts of divine masculine, feminine, and holy family as they appear in other contexts. The Catholic Church emphasizes a relationship with a personal God who is revealed through faith and scripture. To more fully know the divine persons of the Trinity when considered in terms of both the ancient ways and recent occurrences, one might need to re-examine trinitarian thought outside the lens of patriarchy.

Catholic Teachings - A Historical Evolution:

God Transcends Gender: The Catholic Church explicitly teaches that God transcends the human concept of sex and is neither male nor female yet is nearly always referred to in a masculine form. God is understood as pure spirit, infinite, and beyond human limitations, while most descriptors in the bible are masculine.

Language used to describe God, like *"Father,"* is considered metaphorical and not reflective of God's essence being male. The idea of God the Father represents a patriarchal bias in nearly the entire bible that is part of the control narrative that began in the time of the Kings. God's love must be compared to the tenderness of a mother; this is at once both a metaphor and a claim that God has a feminine dimension. Not to prove current teachings wrong, but to prove the current teachings need to be more fully considered in a historical and cultural context so that the doors to greater understanding may be opened.

Focus on the Trinity: The central mystery of Catholic faith is the Trinity of one God in three coequal, co-eternal persons consisting of the Father, Son, and Holy Spirit. This concept of God is misunderstood in terms of the male gender and subordinate family roles that rely on this hierarchy for control and obedience.

Biblical and Historical Context: The Bible, as interpreted by the Catholic Church, uses masculine language to refer to God, though it also employs feminine imagery to describe God's attributes. Historically, societal, and cultural structures in which early Christianity developed were patriarchal, which influenced how divine roles and figures were perceived and emphasized. During this same period of human history, matriarchal roles were minimized and pushed aside as having little to no real meaning as an aspect of God. As Christianity developed within a predominantly patriarchal society, some early Christian thinkers interpreted Jewish traditions and biblical texts in ways that reinforced patriarchal structures within the Church.

Concerns about Syncretism and Idolatry: The Catholic Church emphasizes the unique nature of God as revealed in Christ and seeks to avoid syncretism, which involves blending elements from different religions. Concepts of a divine feminine or goddess figures might be viewed as potentially leading to idolatry or the worship of false idols, which is condemned in the Bible. The divine aspect of God as Mother is not this at all. It is what is now referred to as the Holy Spirit.

The Catholic Church's doctrine of God as a genderless, Triune mystery and its concern for preserving the uniqueness of God's self-revelation in Christ are key reasons why it does not incorporate a *"divine feminine"*. In this way, all aspects of a Triune God are not fully dignified. Might it be worth considering that the Holy Family of Father, Mother, and Holy Son is a closer symbol of the Triune God? A God who loves us all beyond our human understanding. Judaism, as the religious tradition from which Christianity emerged, has significantly influenced Catholic doctrine in several ways. This is particularly true in areas where Judaism has historically demonstrated patriarchal characteristics. The Catholic Church acknowledges the shared spiritual heritage with Judaism and has repudiated supersessionism, which is the belief that Christianity has replaced Judaism as God's covenant people.

In essence, Judaism's historical patriarchal culture has shaped certain aspects of Catholic doctrine, particularly through the adoption of biblical narratives and the use of patriarchal language for God. However, it is essential to understand that Catholic doctrine is also influenced by its unique theological developments and interpretations. Might the time be upon us for a fuller consideration of our shared history, the connections to pre-biblical cultures, and how a more unified approach will make our faith stronger, make our relationship with the creator of the universe more complete in love rather than a need for control and obedience?

Biblical Influences:

Scriptural Foundation: The Catholic Church recognizes the Hebrew Scriptures, or the Old Testament, as divinely inspired and forming a crucial part of its own biblical canon. These texts often describe societal structures that reflect the patriarchal norms of the time, such as men holding positions of authority in families and communities. Men are the high priests and most directly connected to the divine.

Genealogies and Lineage: The Bible emphasizes the lineage of prominent figures, particularly in tracing the lineage of Jesus through the male line, which reflects the importance placed on male descent in a patriarchal society.

Emphasis on Male Leadership: Biblical narratives frequently portray male figures as prophets, priests, and leaders, which has influenced Catholic teachings on male leadership within the Church, such as the reservation of the priesthood to men.

The Divine Feminine within the Holy Trinity:

Critique of Traditional Trinitarianism: This perspective questions the traditional masculine representation of God as Father, Son, and Holy Spirit.

The argument is that it might create an imbalance and potentially diminish the divine feminine. Within the current Christian understanding of God, there is no space for what the ancients called the divine feminine. A return to the divine feminine posits that these diverse phenomena might be unified by the existence and intervention of a divine triad, described as a Father, Mother, and Holy Son, rather than solely by a traditional Father, Son, and Holy Spirit model.

The Holy Spirit as Divine Feminine: Drawing on linguistic and thematic clues, particularly from Old Testament Hebrew, where the word for *"spirit,"* ruach, is feminine, there is a valid argument that the Holy Spirit embodies the feminine attributes of God.

Holy Spirit as Divine Motherhood: This approach suggests a Trinity of Father, Mother as the Holy Spirit, and holy Son, emphasizing the nurturing, compassionate, and life-giving qualities often associated with motherhood as aspects of the Holy Spirit's role within the Holy Trinity.

Theological Arguments and Historical Context:

Linguistic Roots: As mentioned, the Hebrew word for "spirit" is ruach, which is feminine, supporting the idea of a feminine dimension within the divine realm.

Wisdom Literature: The Old Testament's personification of Wisdom is Sophia or Chokma, and both are feminine nouns as a divine figure involved in creation, representing the divine feminine aspect of God.

Early Church and Gnosticism: Some early Christian traditions and Gnostic texts referred to the Holy Spirit with feminine pronouns or equated her with Sophia, a feminine form of divine wisdom.

God and Gender: Re-imagining the Trinity in this way could foster a more balanced view of gender roles within Christianity and provide a more complete image of God, recognizing both masculine and feminine attributes within the divine nature. This reinterpretation is not universally accepted within Christianity.

While traditional Trinitarian theology emphasizes that God is neither male nor female but transcends human gender, the aim of this book is to broaden one's understanding of the divine nature and address historical imbalances in gender representations from Judaism, Catholicism, Protestantism, and all the orthodox subdivisions as well.

The Roots of Judaism:

There are similarities that suggest a common historical event or oral traditions that were adapted, reinterpreted by diverse cultures, and passed down to Semitic peoples, who subsequently have given that same history to Christianity. Others propose that some Biblical narratives may have drawn inspiration from earlier Mesopotamian texts, or that both may have drawn from a common, earlier source. There are several noteworthy correlations between Sumerian mythology and stories found in the Bible, particularly in the book of Genesis.

Creation narratives: Both Sumerian texts, such as the Eridu Genesis, and the Bible describe the creation of the world, including the emergence of dry land from water and the creation of humankind from Earth or clay.

The First Man: In both traditions, the first man is given the responsibility of tending to the animals and caring for the natural world.

Loss of Immortality: There are parallels between the story of Adam and the Sumerian myth of Adapa, where a character's actions lead to the loss of eternal life for humanity, resulting in disease and mortality.

The Great Flood: Both Sumerian and Biblical texts recount a catastrophic flood that devastates the Earth and the salvation of a select few on a large vessel.

Flood Narrative Details: Both accounts include divine decisions to destroy humanity due to its wickedness, instructions to build a large vessel, the use of birds to determine if the floodwater has receded, the vessel resting on a mountain, and sacrifices offered after the flood.

Concepts of Paradise: Both traditions mention a concept of a paradisiacal land, sometimes referred to as *"Edin"* in Sumerian mythology, a place where death did not exist initially.

Creation of Woman: Some scholars point to similarities between the creation of Eve from Adam's rib and the Sumerian myth where the goddess Ninti, whose name can mean *"lady of the rib"* or *"to give life"*, was created to alleviate the god Enki's rib pain.

Why was the Old Testament monotheistic? The emergence of strict monotheism in ancient Israel, as reflected in the Old Testament, was a complex and gradual process rather than a single event. The shift towards a more developed and rigorous monotheism in ancient Israel is depicted in the Old Testament. This development can be viewed as a gradual process influenced by theological revelation, historical events, and ongoing religious and social discourse that is combined with cultural influences of a burgeoning patriarchal society of Semitic peoples.

Contributing Factors:

Early Israelite Religion: While later biblical texts emphasize a clear monotheism, evidence suggests that early Israelite religion was not strictly monotheistic in the modern sense. It may have been closer to henotheism. Henotheism is where Yahweh was worshipped as the chief deity without necessarily denying the existence of other gods.

The Exodus and the Covenant: The Exodus from Egypt and the covenant with Yahweh at Sinai played a foundational role in shaping Israelite identity and religion. It is the interactions on Mount Sinai that some researchers see as ADA experiences for the Semitic peoples, as they were guided by Moses to the promised land.

The Ten Commandments, for example, prohibited the worship of other gods alongside Yahweh, establishing a unique relationship between God and the Israelites.

Prophetic Critique: The prophets played a significant role in challenging syncretism and idolatry, emphasizing the exclusive worship of Yahweh and advocating for social justice. Their messages helped solidify the belief in Yahweh as the one true God.

The Deuteronomistic Reform and Exile: King Josiah's reform movement, followed by the Babylonian Exile, further emphasized the singularity of Yahweh. The Deuteronomistic History, edited during and after the exile, strongly promoted monotheism. During this period, some scholars believe a circle of priests and scribes in exile developed the concept of Yahweh as the sole God of the world.

Influence of Other Religions: While not the direct origin, some scholars believe that contact with monotheistic or monolatrous ideas in Mesopotamia, such as Zoroastrianism, has likely influenced the development of monotheism in ancient Israel. It may be worth it to look more closely at Zoroastrianism, its origins, and history.

In Chapter 15, we have considered how traditional Catholic doctrine on the Trinity has been influenced by historical patriarchal interpretations of biblical and cultural texts, which may have obscured a more complete understanding of God over time. This chapter suggests that the Catholic Church teaches that God is a pure spirit that transcends gender. Those teachings are primarily in the masculine language used to describe God, particularly as *"Father,"* which reflects a patriarchal bias from early Christian and Jewish societies that has likely evolved into God as a masculine being. Chapter 16 then explores the idea that the Holy Spirit might represent the *"divine feminine"* within the Trinity. This idea is supported by linguistic roots, such as the Hebrew word for *"spirit,"* ruach, being feminine, and the personification of Wisdom as Sophia or Chokma in the Old Testament. Chapter 15 also points out that early Christian traditions and Gnostic texts sometimes referred to the Holy Spirit with feminine pronouns. The chapter then proposes that re-imagining the Trinity to include a *"divine feminine"* could provide a more balanced view of God's nature and address historical gender imbalances in religious representation.

Additionally, Chapter 15 draws parallels between biblical narratives, such as the creation story and the Great Flood, and earlier Sumerian myths, suggesting a common historical or oral source that was later reinterpreted. The shift to strict monotheism in the Old Testament is described as a gradual process influenced by theological revelation, historical events, and patriarchal cultural influences.

The chapter argues that this historical and cultural context has shaped Catholic doctrine, particularly through the adoption of biblical narratives and the use of patriarchal language for God. Finally, the history and development of monotheistic thoughts come from Zoroastrianism. Chapter 15's considerations suggest that Zoroastrianism may even be another book to be written later. For the purposes of this work, there is no other consideration of its influences on the culture and society of the Semitic peoples.

Chapter 16
Seers & Saints

The ideas explored throughout the earlier sections have laid the groundwork for a deeper understanding, giving rise to a new frame of reference. Building on these integrated concepts, this chapter turns to the role of seers and saints, examining how their experiences may help bridge the gap between the metaphysical or spiritual and what is understood as the physical world. It is with the hope that through this exploration, a deeper sense of faith may emerge, grounded in a clearer understanding of how these encounters have been seen, interpreted, and recorded. Seers and saints represent a key aspect of the spiritual nature of the divine, one that connects directly to human experience.

The books of Genesis, Exodus, Leviticus, and Numbers all describe an object that was seen by the various peoples and prophets. One such divine being led Moses and his people through the desert for forty years. Ezekiel was visited twice by a flying object that he attempted to describe. David describes visitations in the Psalms where he saw thousands of flying objects. Even Isaiah saw a divine being. Isaiah reported the being as having six wings. Most of the biblical accounts have an association with a type of angel. The two most common types are called cherubim or seraphim throughout the bible. There are some specifics found in nearly all the biblical accounts from Saints and Seers who had no technological knowledge, and so their descriptions of what they saw are in a mythological frame.

They would also use poetic terms in some cases as descriptors of what they saw. Angels are supernatural, spiritual beings, or even what some today might call aliens, depending upon your frame of reference, your faith, and your understanding of how current technology may help describe what was seen, and what we still see today. Many of the biblical accounts speak of the wings and a face of the cherub.

There are several accounts, both in the bible and in the lives of the saints, which show a corresponding theme concerning encounters with the divine. These encounters are not well explained or are described in a poetic manner, causing confusion over what these seers have seen. What was witnessed has been manipulated over time to tell a more unified story for the purposes of those in charge who desire to control the narrative. It may be that those in charge have not meant any harm other than to demonstrate a more unified way to present encounters in a more traditional and acceptable manner. When you strip away the poetic and literary schemes when looking at both biblical, saintly, and modern accounts, what is left are extraordinarily similar occurrences. These similar encounters, visions, and messages all need to be considered in light of current knowledge, sciences, and understanding of technology, especially with the modern occurrence of UAPs and ADAs.

Biblical Appearances - Daniel:

Aided an Unnamed Angel (c. 536 B.C.): Michael is described as *"one of the chief princes"* who came to assist an angel, traditionally thought to be Gabriel, in his conflict with the *"prince of the kingdom of Persia"* (Daniel 10:13).

Stood for the People of Daniel (c. 536 B.C.): He is named "Michael your prince," who "stands guard" over Daniel's people. This passage reinforces his role as a special protector of Israel (Daniel 10:21).

Will Arise for Israel During the End Times: The prophet Daniel is told that *"at that time Michael, the great prince who protects your people, will arise"* (Daniel 12:1).

The New Testament:

Led the War in Heaven: The Book of Revelation describes Michael the Archangel leading an army of angels in a heavenly battle against Satan. Satan is depicted as a dragon, and his angels as demons. This rebellious group of heavenly hosts wanted to take the place of God, the creator of the universe. Satan was called Lucifer, light bearer, and was magnificently beautiful. Michael is victorious and casts the dragon and his followers down to Earth, from Revelation 12:7–9.

The New Jerusalem Study Bible Perspective: The New Jerusalem Study Bible (NJB) is a prominent Catholic translation of the Bible known for its scholarly approach and detailed footnotes and introductions. This work is especially helpful for those who want to gain an understanding of the history and cultural connections of the Sematic peoples to their ancestors from Summer. The notes found in the NJB are key to understanding the historical perspective of where the Old Testament comes from. The introduction to the Pentateuch, in subsection two, explains some of the challenges of the various books of the Pentateuch.

The Introduction to the Pentateuch, in Subsection 2, NJB:

"The progress by archaeologists and historians in reconstructing ancient civilizations adjacent to Israel shows that many of the laws and institutions of the Pentateuch had extra-biblical parallels far earlier than the dates attributed to the documents," and… *"Many of the narratives presuppose conditions other and more ancient than those in which these documents are likely to have been compiled."*

The New Jerusalem Bible (NJB) study edition notes acknowledge significant parallels between the Pentateuch, particularly Genesis, and earlier Mesopotamian literature, including Sumerian and Babylonian texts. The commentary understands these similarities not as direct copying, but as evidence that the biblical authors were aware of and adapted common Ancient Near Eastern (ANE) literary traditions and themes. The NJB commentary addresses the most prominent parallels found between the primeval history in Genesis and earlier Mesopotamian literature, including the Sumerian flood narratives and the Babylonian Epic of Gilgamesh. The notes highlight the cultural similarities and how the Abrahamic lineage comes from among the cultural peoples of Mesopotamia.

Key Sumerian Parallels Discussed in the NJB:

The Flood Narrative: The story of the Great Flood in Genesis shares numerous elements with earlier Sumerian and Babylonian stories, such as the Eridu Genesis and the Epic of Gilgamesh. The NJB notes point out resemblances like:

"A divine decision to destroy humanity with a flood. One righteous man (Noah, Utnapishtim) is chosen to build an ark."

"The survivor is being instructed to take animals aboard. The vessel is coming to rest on a mountain. The survivor releases birds to find dry land."

"A sacrifice offered by the survivor after the flood subsides."

The challenge lies in the fact that the NJB commentary emphasizes how the biblical story reinterprets this tradition from a polytheistic context into a monotheistic one.

Monotheistic vs. Polytheistic: In Mesopotamian accounts, the gods are capricious and squabble among themselves, sometimes sending the flood due to the noise of humanity. In Genesis, God sends the flood as a just judgment for moral wickedness.

Theological Purpose: The biblical account shifts the focus from the arbitrary whims of many gods to the righteous judgment of a single, moral God, providing a theological corrective to the myths of the surrounding cultures. It is of note that both Biglino and Wallace argue that the monotheistic changes found in the Old

Testament are exactly what cause it to be more confusing because the biblical stories are not of a monotheistic God. They are a summary retelling of older stories from the region of Mesopotamia that describe something unknown.

Creation Narratives: While the parallels are less direct with Sumerian creation myths, the NJB notes acknowledge broad similarities in cosmology with Mesopotamian and Egyptian thought. These include the existence of a primordial, chaotic watery state before creation. In Genesis, however, the creation process is portrayed as a deliberate and orderly act of a singular God, in contrast to the polytheistic and chaotic origins found in Mesopotamian myths like the Enuma Elish. Both Wallace and Biglino correct these inconsistencies with a deeper understanding of the word for God or gods and how a proper use renders the biblical accounts as more direct correlations to the stories of the Sumerians.

Genealogical Lists: The genealogies in Genesis 5 and 11 also show parallels with Sumerian King Lists, specifically the ten-generation pattern that appears both before and after the flood.

A primary reason for genealogies is that the life spans pre- and some post-flood lives were counted in hundreds to thousands, to hundreds of thousands of years for both sets of stories. The NJB's commentary notes recognize these structural similarities, viewing them as a formulaic literary pattern common to the region. This pattern is adapted by the biblical authors for their own historical and theological purposes. What needs to be considered is the possibility that they are a summary retelling of the Sumerian stories and, for that reason, have a similar resonance. This certainly raises the question of creation and the Earth as being only around six thousand years old.

Summary of the NJB's Scholarly Approach: The New Jerusalem Bible study edition adopts a historically critical approach to interpreting the Pentateuch. Its notes reflect the scholarly consensus that the biblical authors were a part of the ANE culture and were influenced by its literary and oral traditions. Therefore, the NJB commentary positions the Pentateuch's narratives not as a complete anomaly but as a text that engages with, and re-signifies, the cultural narratives of its time. The NJB sees the biblical account as drawing upon shared traditions while fundamentally reinterpreting them to convey a unique theological message centered on a monotheistic worldview. While this worldview may have served humanity well, there is growing evidence that it may not be exactly how our history unfolded before us.

Douglas Petrovich suggests that having a more fluid point of reference may be required to see the future that is behind us. This is not a misprint, as time seen by Semitic peoples works exactly in this manner, where the past is in the front, and the future is behind when considering a first-person point of view.

The idea of past in front and future behind oneself is a more accurate consideration of time, as the past, through the future, all exist at the same time. This is why today is a gift from God, and why we call today *"The Present."* Douglas Petrovich has a Ph.D., M.A., Th.M., and M.Div., and is a professor of biblical history and exegesis. He earned a Ph.D. from the University of Toronto at the St. George campus.

He has a major in Syro-Palestinian archaeology, a first minor in ancient Egyptian language, hieroglyphics, and a second minor in ancient Near Eastern religions. He is the author of *"The World's Oldest Alphabet."* Dr. Petrovich highlights other examples of differing cultural perspectives, like how distinct cultures perceive time. Most of us view the past as behind us and the future as ahead of us in our modern Western world. The Biblical Hebrew word for *"tomorrow,"* or what westerners understand as the future, is מחר (mahher).

Mahher comes from the root אחר (ahher) meaning "in back." In Biblical Hebrew, the word for *"yesterday"*, or what westerners understand as the past, is תמול (temol). Temol comes from the root מול (mul) meaning *"in front."* The Hebrews of the Old Testament saw time from a place of observance. The Biblical Hebrew frame of reference is, hence, the past is seen and so found in front, and the future, as unseen, is behind the observer. The past is known and seen as in front of the observer. As for the future, to the Old Testament Hebrews, the future is not known and therefore cannot be seen, because the future is not seen, it is found behind the observer. Today, most readers are interpreting Old Testament biblical writings with a Western philosophy that results in mistranslations and misinterpretations of the Old Testament texts. We will have to go into some difficult themes throughout the scriptures and reassemble ideas into postmodern descriptors that are more easily understood in terms of our technology. Modern secular explanations are no longer enough to describe how we can see more clearly what some of the ancients experienced concerning certain types of interactions of many prophets, saints, and seers. It is essential that we learn the philosophy of the Ancient Hebrews to better understand the text we are reading. The Hebrew language is a concrete language with an Eastern philosophical understanding of the world.

What this means is that the Hebrew language uses words that express something that can be seen, touched, smelled, tasted, or heard, and all five of the senses are used to explain things from a frame of reference that uses all five senses. An example of this can be found in Psalms 1:3, where one can read, *"He is like a tree planted by streams of water, which yields its fruit in season, and whose leaf does not wither."*

What is normally called the Old Testament must be considered in terms of who the writers were, when they lived, and what cultural philosophy influenced their lives during the time the Old Testament was written. The perspective of an Eastern philosophy is where the

Hebrews of the Bible wrote the Bible. As the Bible was written from an Eastern philosophical perspective, it is important that we recognize that we cannot interpret it through our own Western philosophy. To do so would place a meaning and interpretation that may not be that of the original authors. Most often, the Old Testament translators translate this way because the original Hebrew makes no sense when literally translated for Western thinkers. We find the concrete sometimes lost to abstract ideas in a passage of a concrete-thinking Hebrew scripture.

This occurs because the words used in translation are abstract English words used to translate the original Hebrew concrete word forms. In his book, *"The Hebrew Genius as Exhibited in the Old Testament,"* George Adam Smith said, *"...the Hebrews were mainly a doing and feeling people. Thus, their language has few abstract terms. Rather, Hebrew may be called primarily a language of the senses. The words originally expressed concrete or material things and movements or actions which struck the senses or started the emotions. Only secondarily and in metaphor could they be used to denote abstract or metaphysical ideas."* The hope is that with this introductory information, one may utilize a more specific frame of reference to be able to find a way to be closer to God and choose the eternal for the greater good of all.

Deuterocanonical and Jewish Apocryphal Appearances:

The following examples are not part of the canon of scripture, and for hardline traditionalists, this is some of the more difficult information to digest. They are important because they were either in large circulation among early Christians or they are specifically and directly referred to in the bible.

Book of Enoch: In this ancient Jewish text, Michael is one of the chief archangels, often described as an angel of justice and mercy who acts on God's behalf. He is also portrayed as a leader among the other archangels.

Assumption of Moses: The Epistle of Jude's reference to Michael's dispute with the devil comes from this apocryphal work, which is not included in the modern biblical canon.

The Epistle of Jude: This epistle recounts a dispute between Michael the archangel and Satan over the body of Moses. Michael is quoted as saying, *"The Lord rebuke you!"* rather than uttering a *"railing accusation"* (Jude 1:9).

Midrashim: Rabbinic tradition includes numerous stories featuring Michael as Israel's advocate. For example, he is said to have saved Abraham from a furnace, protected Sarah from defilement, and replaced Isaac on the altar.

Post-biblical Appearances:

The primary divine being encountered by seers of the church is St. Michael. This divine being has done some rather spectacular things in the name of the kingdom of God. This recount is only some of the interactions from the ancient past.

Monte Gargano, Italy (5th Century): According to tradition, Michael appeared multiple times in a cave on Mount Gargano in southern Italy.

He declared the grotto sacred, and the site became one of the most important pilgrimage destinations in Europe. The first apparition occurred to a bishop, instructing him to build a sanctuary in the cave.

Mont Saint-Michel, France (8th Century): The Archangel Michael is said to have appeared to St. Aubert, the bishop of Avranches, and instructed him to build a church on a tidal island.

After initially ignoring the instruction, Aubert was convinced when Michael pressed his finger into the bishop's skull.

Healing at Chonae, modern-day Turkey (4th Century): Legends recount Michael saving a church at Chonae by channeling rivers underground to prevent a flood. Healing springs associated with the archangel were also venerated at the site.

Ended a plague in Rome around 590 A.D.: During a plague in Rome, Pope Gregory the Great led a procession.

Michael appeared atop Hadrian's mausoleum, sheathing his sword, to signal the end of the pestilence. The mausoleum was renamed Castel Sant'Angelo, or Castle of the Holy Angel.

Saved a Man from Drowning at Mount Athos: An Orthodox tradition from Mount Athos tells of Michael rescuing a young man who was cast into the sea with a stone around his neck by robbers.

In Chapter 16, we have considered that a more fluid point of reference may be required to understand history, and that the past, through the future, all exist at the same time. The chapter summarizes that the New Jerusalem Bible's (NJB) scholarly approach to the Pentateuch, which views the narratives as engaging with and reinterpreting the cultural narratives of the Ancient Near East (ANE), rather than being a complete anomaly. The NJB's commentary sees the biblical accounts as drawing upon shared traditions but fundamentally reinterpreting them to convey a monotheistic message centered on a single, moral God. This is contrasted with the arguments of Biglino and Wallace, who suggest that the monotheistic changes in the Old Testament make the stories more confusing. The confusion, according to these writers, is because the Old Testament was originally a retelling of older Mesopotamian stories that describe something other than a monotheistic God. Chapter 17 also includes an introduction to some of the divine beings who have intervened in human affairs since biblical times.

Chapter 17
Appearances, Apparitions & Flight

As the ideas introduced throughout this book begin to take shape, this chapter moves into a more detailed exploration of the connections between saints, seers, biblical accounts, and the prophetic messages they often herald. Drawing from traditional works such as *Acta Sanctorum*, *The Golden Legend*, and *Liber de apparitione Sancti Michaelis in Monte Gargano*, this section examines how these encounters have been recorded and understood over time. This effort further strengthens the account of St. Michael and his divine role in the lives of the seers.

Old Testament Identifications: Some traditions identify Michael with the unnamed *"angel of the Lord"* who appears throughout the Old Testament. One such example is the one who guided the Israelites during the Exodus.

There are rabbinic traditions that suggest this merkabah, or angel, was the archangel St. Michael. What that gives us in modern terms is a tetrahedron or four-sided geometric triangular shape that also has six edges, so whether four faces and four wings or six wings, a tetrahedron fits nearly perfectly with these biblical accounts. Jewish scholars began to analyze these accounts and eventually began to see the double tetrahedron superimposed one over the other as the Merkabah or flying chariot. In the Book, *"UFOs in the Bible,"* Goudsward likened those terms more closely to a surface or edge when considering the Hebrew words used. A more common name is the Star of David, and this star symbol is the central graphic of the Israeli flag today. It is worth noting that an absolute necessity for space flight as we know it today is a gyroscope. The geometrical qualities of the Merkabah, according to Goudsward, are at least partly gyroscopic in nature. This flying object was also named the hornet by God when He explained to Moses how He would send a messenger ahead of the Israelites to clear the way for them. This object was clearly programmed to perform a protective function to assist the Israelites as a pillar of fire by night and a cloud of smoke by day. This divine intervener has several names:

- *Cherubim*
- *Merkabah*
- *Seraphim*
- *Hornet*

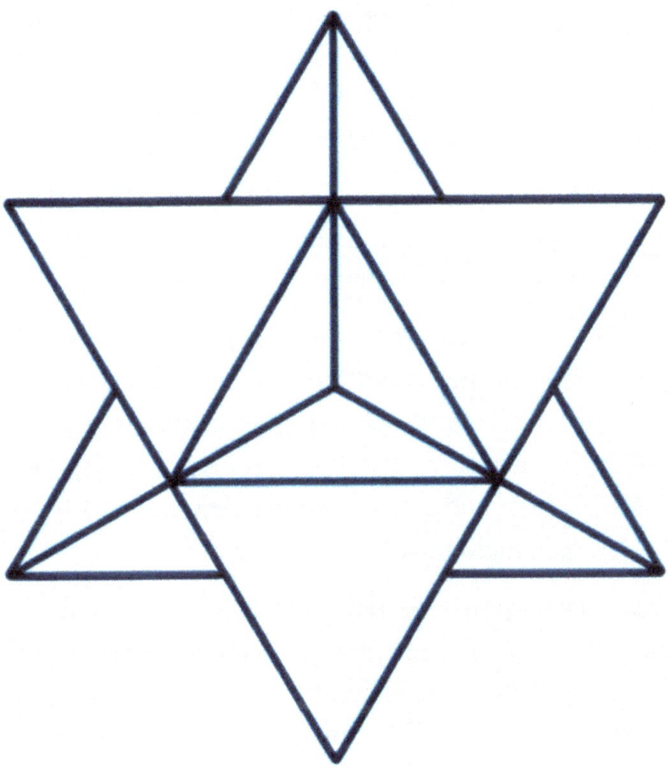

Courtesy: *https://www.dreamstime.com/stock-illustration-merkabah-star-tetrahedron-stellated-octahedron-stella-octangula-can-be-seen-as-d-extension-david-also-called-mer-image41022317*

In more contemporary terms, God as the Programmer seems to have programmed the object to at least be a partially autonomous drone-like craft that was monitored from afar and left to do the bidding of the Programmer, God. This thought will become more important in a later section when we consider connections between science and these objects.

There are several types or at least sizes of these tetrahedron craft, where some carried God along with others like Enoch and Ezekiel to distant places in the encounters of the books of Enoch and in the bible.

Jesus as Michael (Jehovah's Witnesses): The Jehovah's Witnesses' interpretation identifies Michael the archangel with the pre-human Jesus Christ, a belief based on an interpretation of Daniel, 1 Thessalonians, and Revelation.

Islamic Tradition: In Islamic belief, Michael (Mīkāʾīl) is one of the four archangels. He is an angel of mercy who provides sustenance and blessings to humanity.

Various Mythological and Spiritual Traditions: Various mythological and spiritual traditions include concepts of divine triads. While not explicitly described as appearing on Earth as orbs of light or fire, some of these traditions may associate divine beings or energies with light and fire to varying degrees, and in different contexts, some of which include the following:

Hinduism: The Trimurti is the triple deity of supreme divinity, representing the cosmic functions of creation, preservation, and destruction. It consists of Brahma, the creator, Vishnu, the preserver, and Shiva, the destroyer. The Trimurti also represents Earth, Water, and Fire: Earth (Brahma) as the originator of life, Water (Vishnu) as the sustainer, and Fire (Shiva) as the destructor.

The Trimurti concept also symbolizes the three gunas, meaning qualities:

- *Sattva as a goodness, symbolizing harmony by Vishnu*
- *Rajas symbolizing passion and activity by Brahma*
- *Tamas symbolizing darkness and inertia by Shiva*

While they are not explicitly described as visiting Earth as orbs of light or fire, the association with fire and the cyclical nature of their roles aligns with the idea of a transformative, luminous presence.

Celtic Mythology: Brigid is a triple goddess associated with fire, poetry, and prophecy, as well as water, childbirth, the hearth, and healing. She is seen as a keeper of the light and is often pictured holding a flame.

Her flame represents her three aspects, consisting of fire of the hearth, fire of the forge, and fire of inspiration. As a Celtic goddess of fire, Brigid embodies the spirit of light itself, representing spiritual wisdom, intelligence, truth, healing, and prophecy.

Supernatural and Spiritual Beliefs: Spirits as orbs; some people believe that spirit orbs are manifestations of spirits, including angels or disembodied souls, which appear as spheres of light. These orbs can be seen in photographs or even with the naked eye. Orbs are sometimes believed to be a simpler form of manifestation when spirits lack enough energy to appear in full body apparition. Distinct colors of orbs are sometimes associated with distinct types of angelic energy or intentions. While the Hindu Trimurti and Celtic goddess Brigid are associated with light and fire, and modern supernatural beliefs link orbs of light to spirits or angels, there is only one known account that explicitly details a divine triad visiting Earth. This is the account of Chris Bledsoe in his book *"UFO of God."*

Angels - Incorporeal Spiritual Beings: Any physical form they take is a manifestation to serve a specific purpose, such as delivering a message or intervening in a crisis. In visions, angels can project images that are recognizable to the person receiving the vision, such as a traditional angelic form, or may appear as a less-defined energy, like a ball of light.

The Arc Angel St. Michael: The shape and appearance of St. Michael in art and apparitions are symbolic ways to represent his divine power and protective role, rather than reflecting a single, definitive physical form. Michael the Archangel is a revered figure who appears in the sacred texts of Judaism, Christianity, and Islam, as well as in apocryphal traditions and religious folklore. His appearances are not a single event, but a cumulative story drawn from various religious narratives spanning centuries. St. Michael the Archangel has appeared to several saints throughout the Church's history, often to deliver a message or provide aid during a time of struggle. Notable accounts include his appearances to St. Joan of Arc, St. Francis of Assisi, and St. Aubert of Avranches.

St. Joan of Arc (c. 1412–1431): Beginning around 1425, when she was 13 years old, St. Michael the Archangel was the first of the saints to appear to St. Joan of Arc. His messages instructed her to be a good child and to live a holy life. As she grew older, he revealed her divine mission to lead the French army against the English invasion and ensure the coronation of Charles VII.

1425: Michael first appeared to Joan while she was in her father's garden, surrounded by bright light. He told her not to be afraid and instructed her to lead a virtuous life.

Gradual Guidance: Over the next few years, Michael appeared to Joan multiple times, along with St. Catherine and St. Margaret. Together, they guided her in her spiritual and physical preparations for her mission.

During her imprisonment, even after her capture, Michael continued to visit Joan, encouraging her as she faced her trial and martyrdom.

St. Francis of Assisi (c. 1181–1226): St. Michael's influence on St. Francis is tied to a penitential period known as "St. Michael's Lent."

1224: St. Francis observed a special 40-day fast in preparation for the Feast of St. Michael on September 29. During this time, while praying on Mount La Verna, St. Francis received the stigmata, the wounds of Christ, from a six-winged seraph. This descriptor of the angel was considered in the discussion of the merkabah earlier in this section. According to tradition, this vision was intricately connected to his devotion to St. Michael.

St. Aubert, Bishop of Avranches (early 8th century): St. Michael appeared to Bishop Aubert, instructing him to build a sanctuary in his honor.

708 A.D.: St. Michael appeared to Aubert three times, commanding him to build a church on the tidal island that would later become known as Mont Saint-Michel in Normandy, France.

St. Gregory the Great (c. 540–604): St. Michael's appearance to Pope Gregory is connected to the cessation of a plague in Rome.

590 A.D.: During a penitential procession, Pope Gregory the Great saw St. Michael above the mausoleum of Hadrian, sheathing his sword; this mausoleum is now the Castel Sant'Angelo.

This vision signaled the end of a devastating plague, and that is why the castle is now named in the archangel's honor.

St. Lorenzo Maiorano, Bishop of Siponto (5th century): St. Michael appeared several times to Bishop Lorenzo of Siponto on Mount Gargano in Italy, a site that became one of the most famous pilgrimage destinations for the archangel.

490 A.D.: St. Michael appeared to Bishop Maiorano, revealing that a cave on the mountain was sacred to him. He instructed the bishop to consecrate the grotto.

492 A.D.: After a military defeat, Bishop Maiorano fasted and prayed for three days. St. Michael appeared to him again and promised victory, which was subsequently achieved.

493 A.D.: Michael returned to tell the bishop that he had consecrated the grotto himself, and therefore no human consecration was needed.

Diego Lázaro (17th century): St. Michael also appeared to a young indigenous convert named Diego Lázaro in Mexico, whose vision led to a new shrine.

1631: St. Michael appeared to Diego and pointed out a miraculous healing spring in a ravine in Tlaxcala, Mexico. He asked that the people be informed of the spring.

Secularized Divinity:

In recent years, the secular terminology concerning divine flying craft has shifted from the religious to techno, scientific, or secular explanation. What was once angels and divine visitations is now *"Unidentified Flying Objects"* (UFOs), which have evolved into *"Unidentified Anomalous Phenomena"* (UAPs), and most recently morphed into *"All Domain Anomalies"* (ADA). These naming changes are thanks in large part to governments and military terminology that are attempts to both delegitimize and legitimize these occurrences because they want to control the narrative.

Because there are also several reverse-engineered crafts, many of which are unacknowledged, the military industrial complex has been accused of planning a false flag. A false flag is where an extra-terrestrial invasion will be publicized to take control of the globe. The terms UFO, UAP, and ADA are all designed to encompass objects observed in the air, sea, or space that cannot be immediately identified.

There is a direct correlation to the previous discussions and what we see and encounter today, using technological explanations rather than religious or poetic ones, because we, in the modern age, have a vastly different frame of reference based on technology and science. We must reframe our understanding of at least the more difficult ADAs as a divine intervention, and the next human awakening is brought to us through ADA experiences that accelerate spirituality, consciousness, and a relationship with our creator that is deeper than ever before. The descriptions of these more difficult to explain ADAs, reported in modern times, often include the following characteristics:

Shapes: Commonly reported shapes include spheres, orbs, triangles, discs, and cylinders. Less commonly mentioned are boomerang, V-shaped, diamond or rhombus-shaped, egg or oval-shaped, rectangular, or box-shaped, teardrop-shaped, or even amorphous or indistinct shapes.

Size: Reported sizes can vary from a few feet in diameter to the size of a tennis court, football field, or even larger.

Colors and Materials: UAPs are frequently described as being white, silver, metallic, or translucent. Some are reported as glowing or emitting light.

Movement and Behavior: ADAs have reportedly exhibited anomalous flight characteristics, including:

- High supersonic speeds
- Sudden accelerations
- Sharp changes in direction, including right-angle turns
- Hovering motionless for extended periods
- Moving against the wind without apparent propulsion

Trans-medium: The term "trans-medium" specifically refers to objects or devices that can operate across different mediums. It is this capability that has given rise to the term *"All Domain Anomaly"* (ADA).

Physical mediums include air, water, and space, without displaying characteristics readily understood by sensors, observers, or our current understanding of physics, flight, fluid dynamics, and space travel. Trans-medium ADAs can move between air, water, and outer space without hindering performance from friction or drag. Some reports indicate that these ADAs lack visible means of propulsion, such as wings, jet engines, or exhaust trails, and are described as silent, while other ADAs emit a low-frequency hum. The Department of Defense considers ADAs as anomalous detections in one or more physical domains, like airborne, seaborne, spaceborne, and/or trans-medium, that are not yet attributable to known actors and that demonstrate behaviors that are not readily explained by our current understanding of the sciences.

Most Common ADA Descriptions:

Orbs: These objects vary in size and often display high speeds and maneuverability. Witnesses have sometimes described them as glowing balls of light. The word "Orb" is most used to describe something that is spherical or shaped like a ball.

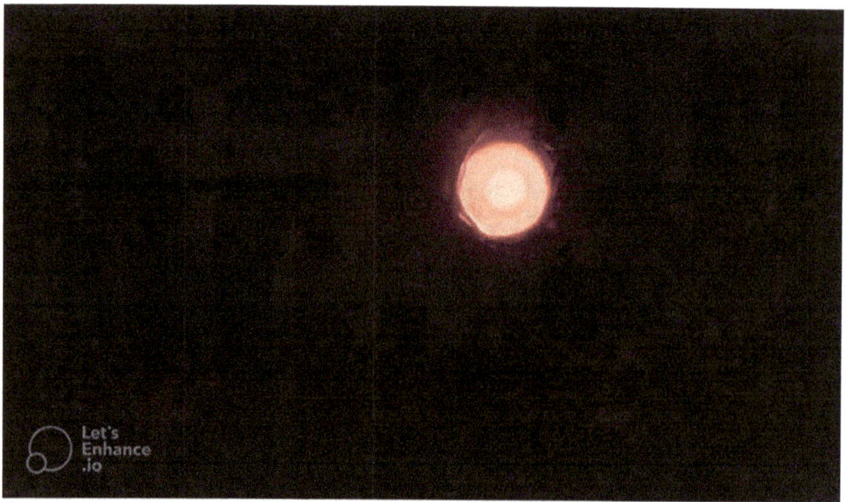

Courtesy: https://i2-prod.mirror.co.uk/incoming/article8604505.ece/ALTERNATES/s615b/UFO-Paris.jpg

It is often used poetically or in more formal contexts. Here are some common ways to describe orbs, depending on the context:

Sphere	Ball
Globe	Small, circular light anomalies
Circular	Translucent
Semi-transparent circles	Glowing light
Radiant objects	Floating

Having a particular color (e.g., white, blue, green, red, yellow, purple, black).

Appearing solid from the center, moving quickly.

Having intricate, multicolored interior designs.

Moving with intention and direction, sometimes leaving trails of light. Moving like a mouse on a keyboard is a way to describe the lightness of the movements.

A common way to describe interdimensional orbs is to consider them as manifestations of entities or energy forms from other dimensions or planes of existence that are often described as:

- *Small, circular light anomalies:* They appear as spheres of light in photographs and videos.
- *Varying in size, brightness, and color:* They can be as small as a crumb or as large as a basketball and can appear in a wide range of colors.
- *Indicative of spiritual presence:* Some believe they are the simplest form of a spirit's manifestation, a way for spirits or angels to make their presence known or travel.
- *Related to portals or gateways*: Some theories suggest they might even be portals that create interdimensional gateways.

- *Often seen in locations with suspected supernatural activity:* Their appearance can be linked to areas with reported supernatural phenomena.

Courtesy: (Altered Photo). By: Twitter/Chip, and enhanced by let's enhance.io, https://knewz.com/tic-tac-ufo-with-attachment-at-back-spotted-hovering-above-city-in-daylight/

Tic Tac: These objects are oval, pill-shaped, or like an egg on its side, which some have reported as similar to a breath mint. A well-known example is the *"Tic Tac"* incident in 2004, where Navy pilots encountered a 45-foot object off the coast of Southern California.

They are sometimes described as smooth and white or silver. Reports often highlight their rapid movement, including quick changes in direction and acceleration.

Spheres with Cubes Inside: This ADA appears as a dark cube inside a clear or translucent sphere. This shape was reported by Navy pilots flying off the East Coast of the United States near Virginia.

Flying Pyramids: These UAPs are pyramid shaped. A video of such objects was reportedly taken by a Navy pilot in 2019 and confirmed by the Pentagon as an authentic image of an ADA.

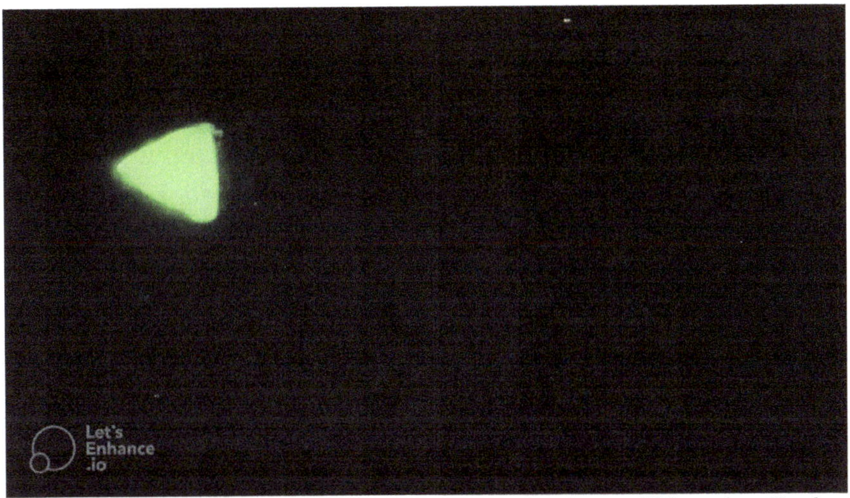

Courtesy: An image of a UAP, seen through night vision goggles, shown to US lawmakers, Naval Air System Command, and enhanced by Let's enhance.io; https://www.bbc.com/news/world-us-canada-64252340

To better understand what the pyramid is, one must consider the Johnson Solid, Square Pyramid, or an equilateral square pyramid, where all the edges are of equal length. These are convex polyhedral with regular polygon faces, but not necessarily uniform like Platonic solids. However, some experts suggest these could be explained by known aircraft or radar-reflecting balloons. The challenge is that many of the ADAs seen today simply cannot be explained with our current understanding of flight and physics.

Chapter 17 discusses figures such as Michael the Archangel, the Hindu Trimurti, and the Celtic goddess Brigid, noting their association with concepts like light, fire, creation, and destruction. It details specific accounts of St. Michael's appearances to various saints, often involving bright light or visions. Chapter 17 introduces the belief that spirits or angels can manifest as *"orbs of light"* when they lack the energy to appear in full physical form. These orbs are often described as small, circular light anomalies that can vary in color, size, and brightness. There are several ADAs left out of this discussion intentionally because of suspicions of reverse engineering human craft that are unacknowledged. There will be a more in-depth discussion of this in a later chapter.

Secular Terminology and ADAs: The chapter highlights a direct correlation between these older divine visitations, modern *"Unidentified Anomalous Phenomena"* (UAPs), and *"All Domian Anomalies"* (ADAs). In this supposed post-Christian, modern era, the terminology has shifted from one of faith and religion to techno scientific. Most modern ADAs are frequently described as having various shapes like orbs, spheres, triangles, discs, that include colors from white, silver, to metallic, and glowing, which also exhibit anomalous flight characteristics like high speeds, sudden accelerations, and the ability to move through different mediums such as air, water, and space.

The most common descriptions of ADAs include *"orbs," "tic tac," "spheres with cubes inside,"* and *"flying pyramids."* In Chapter 17, we have considered a shift in how unexplained phenomena are perceived. Historically, divine visitations and angelic appearances, such as those of Michael the Archangel to various saints, were interpreted through a religious lens. This chapter explores the evolution of how humans interpret and describe unexplained phenomena, shifting from religious and mythological explanations to modern, secular terminology. In modern times, similar occurrences are often secularized and categorized as ADAs by governments and the military, using technological and scientific terminology. This chapter suggests a direct correlation between historical religious interpretations and modern technological explanations for these events.

Chapter 17 proposes that a deeper understanding of ADAs may lead to human awakening, accelerating spirituality, and a relationship with the Creator of the universe. Several potential conclusions can be drawn by applying the CTMU from Part II to UAPs here in Part III, along with biblical accounts, mystical experiences as manifestations of the divine, and messages of the coming solar nova. The CTMU emphasizes the interconnectedness of mind and reality that challenges the traditional separation of scientific anomalies (ADAs), religious narratives like biblical encounters, and subjective mysticism like saintly experiences, and their messages of a coming solar nova. The CTMU suggests a unifying principle or structure underlying these phenomena, consistent with the synthesis arguing for a continuous interaction with the divine.

Chapter 18
So, What is ADA Anyway?

As humanity continues attempting to understand the luminous phenomena witnessed across history, the language used to describe them shapes how they are interpreted. Terms such as UFO, UAP, and ADA often carry modern governmental or military implications that may fail to reflect the deeper spiritual and historical significance many associate with these manifestations. To move beyond those limitations, this chapter proposes a series of new descriptors rooted in classical etymology that better capture the fiery, radiant, pulsating, trans-medium objects historically linked to angelic and supernatural visitations. These terms are not intended as final definitions, but rather as a framework for expanding understanding and creating a new reference point that more closely reflects the divine nature attributed to these phenomena.

Transluminuspyrorb: A transluminuspyrorb is a supernatural object that evokes a fiery sphere emitting and transmitting light and energy through various mediums, while also suggesting that it has a radiant and pulsating nature. A transluminous orb is any ball-shaped object that may also be somewhat egg-shaped and can be described as oval or ovoid. Think of it like a circle that has been stretched out, so it is a bit longer in one direction than it is wide.

Root Etymologies:

Pyrorb: From the Greek root *"pyr"* meaning *"fire"*, combined with *"orb"*, which originates from Latin *"orbis"* meaning *"circle, disk"*. This ending in the name immediately conjures the idea of a fiery, spherical object. This also evokes the image of radiant, piercing, glowing, pulsating objects.

Trans: This Latin prefix means *"across"*, *"beyond"*, or *"through"*. It captures the *"transmedium"* and *"transparent"* qualities, implying the ability to move through or be seen through different substances.

Luminus: Derived from the Latin *"lumen"* meaning *"light"* and reflecting *"luminous"*. This signifies the *"light"*, *"radiant"*, and *"glowing"* attributes of the object.

The transluminous orb can also be like an egg because it is rounded, but a little pointy or narrower at one end and wider at the other. Imagine squeezing a perfect ball a little bit, and that's kind of the shape you would get. Because everything needs to be categorized in human terms, there are sub-groups of craft as follows:

Pyro-Transluminus Ovoid (PTO): A Straightforward combination, suggesting a ball-like structure that emits light, can pulsate, sometimes appears as fire, and is able to operate in a transmedium fashion.

Pyro-Transluminus, Cubidocentric Orb (PTCO): A Straightforward combination, suggesting a cube in the center of an orb-like structure that emits light, can pulsate, sometimes appears as fire, and is able to operate in a transmedium fashion.

Pyro-Transluminus, Pyramidocentric Orb (PTPO): A Straightforward combination, suggesting a pyramid in the center of an orb-like structure that emits light, can pulsate, sometimes appears as fire, and is able to operate in a transmedium fashion.

Pyro-Transluminus, Merkahmid (PTM): A straightforward combination, suggesting a pyramid-like structure embodying the energetic and physical principles of the Merkabah that emits light, can pulsate, sometimes appears as fire, and is able to operate in a transmedium fashion. This is by no means an exhaustive list or mutually exclusive. These categories simply help us place the various crafts in familiar groupings for discussion purposes. It is likely that they are all the same thing that just manifests differently to different people based on their ability to see and understand what is seen.

In basic geometry, an ellipsoid is a surface where every plane section that cuts through it forms an ellipse or a circle. An ellipsoid is defined by three semi-axes or radii along perpendicular directions. *"Ellipsoidal"* relates to an ellipsoid, which is a three-dimensional geometric shape resembling a sphere that has been stretched or squashed. In simple terms, you can think of an ellipsoid as a *"squished"* or *"stretched"* sphere.

The Earth is a prime example of an oblate ellipsoid, bulging slightly at the equator due to its rotation. These unfamiliar terms are modern descriptors for a transcendent origin point or a state of pure potential, containing and embodying the diverse aspects of existence in a delicate and otherworldly form, possibly radiating light or energy, and serving as the source of a multidimensional reality. Considering the myths and mythological motifs around creation, where the cosmic egg represents primordial potential, a "multi-modal Celestial Luminous Ethereal Ovum" could refer to UAPs as entities with these qualities:

Multi-Modal Celestial Luminous Ethereal Ovum (M2-CLEO): The short form M2-CLEO is not a standard scientific or philosophical term.

The meaning can be understood by analyzing each word:

Multi-modal: Indicates multiple modes of communication or data processing. In AI and technology, it refers to systems that integrate data from text, images, audio, and video.

Multiple aspects or forms: The *"multi-modal"* aspect could signify that it represents or contains diverse forms, levels, or dimensions of existence, or interacts through various channels, embodying the interplay of different modes of being in that same way that the military classifies some UAP as transmedium.

Transmedium: This term, particularly relevant in the context of Unidentified Anomalous Phenomena, and All Domain Anomalies (UAPs, ADAs, or formerly UFOs), describes objects capable of moving through different environments like air, space, and water.

Celestial: This refers to things relating to the sky or outer space, often associated with astronomy and the stars. Can also relate to the heavens, or something heavenly, divine, or spiritual.

Luminous: Meaning is a thing that is emitting or reflecting light, shining, or glowing. Figuratively, it can also describe something clear, easily understood, enlightening, or wise.

Ethereal: Meaning delicate, light, airy, or otherworldly. It can also refer to something spiritual, lacking material substance, or heavenly.

In ancient Greek philosophy, it was used to describe the pure substance or ether believed to compose the heavens. This implies a light, delicate, and often non-physical or spiritual quality.

Ovum: Represents potentiality, the beginning of life, and the containment of genetic information. In biology, an ovum is a female reproductive cell or egg.

In the divine sense, it is pure; it is the creative force of the creator with the following attributes:

Source of creation: Like the cosmic egg in myths, it would be the origin point for a vast and complex entity, such as a universe or a reality where divinity comes to the corporeal.

Connection to the heavens/divinity: Being *"celestial"* and *"ethereal"* would link it to a spiritual or otherworldly dimension.

Radiating light or energy: "Luminous" would imply it emanates light, energy, or knowledge.

Delicate and otherworldly: Its ethereal nature would suggest it is not a tangible, material object, but rather something subtle, delicate, and possibly spiritual.

What Does It All Mean?

Based on these combined concepts, a Multi-Modal Celestial Luminous Ethereal Ovum (M2-CLEO) could potentially be interpreted in a few ways, depending on the context and beliefs, as follows:

The Unified Concept: An M2-CLEO is an entity that is advanced or otherworldly and represents a real portion of the current UAP phenomena. The M2-CLEO are divine beings or objects from heaven that transcend our current understanding of physics, biology, and the quantum realm.

Metaphorical or Symbolic: The term M2-CLEO could be used metaphorically in artistic, spiritual, or philosophical contexts to represent something like:

The Origin or Essence of Life in a Cosmic Sense: A *"celestial ovum"* could symbolize the birth or beginning of life on a grand scale, perhaps related to theories about panspermia or cosmic genesis, or a reference to the object's origin as a place of divinity.

A Spiritual or Energetic Entity: Given the use of *"ethereal,"* The M2-CLEO represents a spiritual or non-physical being or energy form.

A Creative Force or Idea: The M2-CLEO could be used in a metaphorical sense to represent the manifestation of a new way to understand an age-old intersection of the corporeal with the eternal.

Chapter 18 proposes updated terms for describing Unidentified Anomalous Phenomena (UAPs) and All Domain Anomalies (ADAs) to shift the discourse away from government and military terminology. The author suggests novel words like Transluminuspyrorb, which is a supernatural object described as a fiery sphere that emits and transmits light and energy. This chapter also proposes sub-categories like the Pyro-Transluminus Ovoid (PTO), the Pyro-Transluminus, Cubidocentric Orb (PTCO), and the Pyro-Transluminus, Pyramidocentric Orb (PTPO), based on their shape and ability to move in a transmedium fashion.

Chapter 18 introduces the term Multi-modal Celestial Luminous Ethereal Ovum (M2-CLEO) to describe some ADAs as entities with qualities of being multi-modal, celestial, luminous, ethereal, and ovum-like. The term M2-CLEO is used to suggest that these phenomena are most likely divine beings or objects from a transcendent origin point, meaning that they are multidimensional. The ovum part of the term is used metaphorically to represent the beginning of life in a cosmic sense or a creative force. The overall conclusion is that these updated terms provide a new frame of reference for understanding these ADAs as supernatural visitations that embody the diverse aspects of existence and a multidimensional reality.

Wrap Up:

When considering the totality of this book, you find an interconnected web that is a matrix of information from across human history to give humanity opportunities to choose salvation, love, and beauty over hate, fear, and control.

Spirituality and Anomalous Phenomena challenges traditional separations by proposing that scientific anomalies such as ADAs, along with religious narratives like biblical accounts, and subjective mystical experiences like those of Edgar Cayce, Chris Bledsoe, the saints, and modern seers are all manifestations of the same underlying unifying principle described by the CTMU. The more things change, the more they stay the same, and our human condition is one of deep-seated need for the creator, for a connection with creation, and a connection to our creator. It is all these various divine interventions over time that have led to greater understanding and ability to find solutions that benefit all of humanity for those with the open ear, or rather open heart, to hear. With this new framework, it becomes clear that there is a purpose and the divine is noble, love, beauty, and the creator has an interest in our well-being. It is so much so that this new frame of reference opens the doors of possibility that allow us to consider the greater existential threats that loom on our horizon.

The challenge then becomes how we might, as humanity, be better prepared to deal with any existential threats, as they are natural occurrences and a part of the creation in which we live. As you take in all that Part I has offered, you will find that this book is a means to move our understanding to the cycles, and science involved in these various processes that are not easily understood in human terms. The processes found in the universe concerning free will, evil, God, the apparitions, M2-CLEOs, visions, and prophetic messages all fall into a mechanism that was described in Part II titled *"The Science."*

In Part III, the focus turns specifically on both galactic and solar cycles as the actual purpose for the apparitions, M2-CLEOs, visions, and prophetic messages that will be discussed in more detail in subsequent chapters. Part II, *"The Cycles,"* considers how to see the bigger picture of nature at the level of stars and galaxies, so that a greater understanding of these processes will be seen as a driving force for all of humanity, both in the near term and long term. It is this unified approach coming from a divine perspective that will be most assistive to our survival into the future. For many, the only way is to prepare the heart and soul for whatever comes because of these natural repeating cycles.

Chapter 19
Disclosure

As the ideas developed throughout the earlier chapters begin to converge, the subject of disclosure becomes increasingly central to the broader discussion. Public interest in UFOs, ADAs, and non-human intelligence (NHI) has grown substantially in recent years, bringing renewed attention to claims of suppressed knowledge and covert government activity. Within this context, Dr. Steven Greer's concept of asymmetric disclosure offers a framework for understanding how information related to modern encounters, reverse-engineered craft, and alleged military-industrial secrecy may gradually become known. This chapter examines those claims and their implications within the larger narrative surrounding disclosure.

Key Aspects of Greer's Asymmetric Disclosure Plans:

Exposing suppressed truths: Greer aims to bring to light military industrial secrets that cover decades of concealed information. He asserts that governments, militaries, and covert or black programs have actively hidden evidence of extraterrestrial life and advanced technologies. He believes that military contractors have and are developing innovative technology from this advanced otherworldly technology.

Corroborating Ideas: Greer's ideas are corroborated by the testimony of the person widely considered to be the *"Groom Lake whistleblower"*, Bob Lazar. In 1989, Bob Lazar publicly claimed to have worked on a top-secret government project to reverse-engineer extraterrestrial technology at a site called S-4, located south of the main Area 51 facility near Groom Lake, Nevada.

Key Details of Lazar's Story:

Initial Interview: Bob Lazar first appeared anonymously in a 1989 interview with investigative journalist George Knapp on Las Vegas station KLAS-TV, initially using the pseudonym *"Dennis."* Dennis spoke of his work, the craft he was attempting to reverse engineer, and the implications of this modern technology on a global scale. He later revealed that he was not Dennis; however, he was Bob Lazar.

Revealing Reverse-engineered Technology: He claims that powerful interests suppress knowledge of advanced, potentially game-changing technologies, potentially derived from extraterrestrial sources, to protect their control over geopolitical systems and economic structures based on fossil fuels and other industries. Like Greer, Lazar alleges that while at S-4, he saw several flying saucers of alien origin and participated in efforts to understand their propulsion systems. He claimed that these UAP were powered by a stable element known as Element 115.

Urging Public Disclosure and Transparency: Greer advocates transparency and the public release of information related to UFOs and extraterrestrial contact, believing that this disclosure could unite humanity, address global challenges, and lead to a new era of technological advancement and peace. Essentially, Greer's approach emphasizes the need to expose the "hidden truths" behind decades of both military and government secrecy and disinformation surrounding UFOs, ADAs, and the potential impact of advanced technologies on our planet and civilization. Other individuals are sometimes associated with Groom Lake claims, but Bob Lazar is the most well-known *"whistleblower"* regarding extraterrestrial technology at the site. His 1989 interviews were instrumental in launching the modern public interest in Area 51, UFOs, and, more recently, in ADAs.

Greer goes on to argue that the military industrial complex has abused its authority by not specifically answering to other branches of government oversight for nearly seventy years, which began during World War II in the mid-1940s. He goes as far as to say that many black budget projects are operating outside of the Constitution in an entirely illegal manner and must be shut down with perpetrators held accountable. At the same time, these claims are extremely hard to substantiate under current anti-terrorism laws in a post-September 11th world. There is hope that the greater good for all of humanity will prevail, and truth with full disclosure will rule the day. There are increasing numbers of whistleblowers who want out, want the truth to be known, and are hoping for a more public discourse on how to proceed with this new knowledge concerning UAP and reverse-engineered technology.

Whistleblower Options: If you are among those who might like to bring specifics to light, or you know someone who is struggling with what to do. Here are some options:

Witnesses: witnesses@siriusdisclosure.com

DP Legal Team: disclosure.legal@gmail.com

www.dni.gov

https://www.dni.gov/ICIG-Whistleblower/index.html

https://www.dni.gov/ICIGWhistleblower/protected.html#hotlines

Other Ways of Disclosure:

"Invisible College": According to Diana Pasulka, there is an *"invisible college"* in academia and research. An *"invisible college"* refers to informal networks of researchers who collaborate, share ideas, and influence scientific discourse outside of formal institutional structures.

Contribution to Disclosure: Members of these networks can play a role in informal knowledge sharing, which, while not a formal disclosure, can shape understanding and pave the way for eventual formal dissemination of information through publications and conferences.

Specifically, an *"invisible school of knowledge,"* as explained by Pasulka, can be understood as the growing understanding and awareness surrounding the experiences of individuals in industry, sciences, and academia that are not immediately apparent.

The "Disclosure Movement": Concerning UFOs, Non-Human Intelligence (NHI), and UAP. In this specific context, "disclosure" refers to the demand for governments to release alleged secret information about ADA, UFOs, and *"non-human intelligence"* (NHI).

Adherents of the *"disclosure movement"* believe that a sudden revelation of hidden knowledge by governments would bring about positive societal change. It is important to differentiate between the various interpretations of *"invisible school of knowledge"* and *"disclosure."*

Each interpretation highlights an unusual way in which knowledge, whether formal, informal, or even speculative, can contribute to the sharing of information, be it within specialized communities, for social awareness, or concerning matters of public speculation.

Bledsoe's Disclosure:

Chris Bledsoe claims that a figure he refers to as *"The Lady"* informed him about an alignment that will occur in April 2026, around Easter. A vision from Chris Bledsoe is mentioned, where the sun is seen growing larger and blowing a portion of the Pacific Ocean into space. These commonalities suggest this could be both a past and a future event. Part II of this book shows the links of Bledsoe's account to Carlson's theory by suggesting that the frozen ocean water blown into space eventually falls back to Earth, causing extinction-level impact events.

According to Bledsoe, The Lady explained that the star Regulus will appear on the horizon in front of the Sphinx, and it will be red. This event is said to mark a shift into new knowledge. It suggests a new understanding is coming to humanity. Bledsoe himself notes that he was not sure what this meant initially but has come to understand that this period around Easter 2026 represents *"the end of an old way and the beginning of a new."* He emphasizes that this is not an apocalyptic event, and it does not signify the return of Jesus, but rather the dawning of new knowledge. There is no astrological or astronomical basis for Regulus to rise *"red"* at the Sphinx soon. Regulus is typically a blue-white star. However, some sources associate the rising of Regulus with the Sphinx in a metaphorical or prophetic context. The Lady and Easter 2026 suggest an awakening to new knowledge. According to Bledsoe, *"The Lady"* is a radiant, divine feminine presence.

For those who doubt and have skepticism, it causes a break in the spiritual connections she gives. Grounding in love, not fear, is the key to connecting with her. Chris reports that he was taken by *"The Lady"* and other unknown human-like entities that seem to be her assistants, and he was shown the sun from the outer atmosphere of Earth. The Sun became so large that it began to overtake the sky and the Pacific Ocean, both evaporating and being thrown out into space. While Bledsoe was not sure if this vision was a past or a future event, we will surmise in these sections that it is cyclical and recurs on several cycles over time and across space.

Current Understanding of Regulus:

Regulus is a blue-white star, not red. Although it will eventually become a red giant star in the distant future, which is estimated in millions of years, this change in color is part of its stellar evolution and not a cyclical event that would align with specific dates in the context of the Sphinx. Astronomical alignments are real phenomena, and ancient Egyptians were skilled in observing and aligning structures with celestial bodies. However, the specific interpretations and prophecies linked to Regulus rising red at the Sphinx are speculative and not supported by astronomical observations alone. Conjunctions, where celestial bodies appear close together in the sky, do occur. For example, Mars, which is red, and Regulus had a close conjunction on June 17, 2025. This presented a visual contrast between Mars' red glow and Regulus' blue-white light. However, this is distinct from Regulus itself, which appears red. Based on astronomical calculations and the Sphinx's Easterly orientation, Regulus will rise facing the Sphinx around the following dates:

- *Early September:* The heliacal rising and first appearance before sunrise of Regulus generally occurs during the first week of September.
- *Around the September Equinox:* Currently, Regulus is visible in the eastern sky before sunrise.

It is important to note that due to the Earth's precession and slight variations in the timing of astronomical events, the exact date may shift slightly from year to year. Regulus is typically obscured from view for about a month, from late in August to early in September, when it is in conjunction with the sun. While astronomically it is not plausible for Regulus to rise red at easter in 2026, there are several atmospheric conditions that would make it favorable, possible, and plausible to rise red in September2026.

Possible Ways to Change the Color of a Star: The Earth's atmosphere can change the apparent color of a star through three main processes: atmospheric extinction, atmospheric dispersion, and scintillation. These effects are more pronounced when a star is observed closer to the horizon, where its light must travel through a thicker layer of the atmosphere.

Atmospheric Extinction: Atmospheric extinction is the dimming and reddening of starlight as it passes through the atmosphere. This is due to the absorption and scattering of light by air molecules and dust particles.

Wavelength Dependence: This scattering is highly dependent on the wavelength or color of the light coming from a particular star. Shorter wavelengths, such as blue and violet, are scattered more efficiently than longer red wavelengths. This is the same effect that makes our daytime sky blue.

Resulting Color Change: As a star's light travels through the atmosphere, more of its blue light is scattered away. The remaining light is enriched with longer, redder wavelengths, causing the star to appear slightly redder than it would from space.

Atmospheric Dispersion: Atmospheric dispersion occurs because the atmosphere acts like a prism, separating a star's light into its component colors. This happens because the refractive index of air is different for different wavelengths of light.

Prism Effect: The atmosphere bends or refracts starlight, and blue light is bent more than red light. As a result, when a star is near the horizon, its image is smeared into a tiny vertical spectrum, with the blue end appearing higher in the sky and the red end lower.

Resulting Color Change: While usually not noticeable to the naked eye, this effect is easily visible in a telescope as a colored fringe on a bright star. With a highly turbulent atmosphere, this dispersion can cause a star to appear to flicker rapidly through a series of distinct colors.

Scintillation or Twinkling: Scintillation is the rapid change in a star's apparent brightness and position due to atmospheric turbulence.

Atmospheric Turbulence: Pockets of air with different temperatures and densities constantly move across the line of sight to a star.

These air pockets bend and distort the starlight, causing its image to blur and shimmer.

Resulting Color Change: Because distinct colors of light are bent at slightly different angles of dispersion, the twinkling can cause rapid, momentary color shifts as the star's light passes through the moving turbulence. This effect is most noticeable when a star is low in the sky, as its light travels through more turbulent air.

The most important note here is that most prophetic visions are not very precise on dates, and for good reasons. So, while it is possible that Regulus could rise red at easter, there is also a chance that it may not happen until later in the year as well. The science demonstrates that there are mechanisms that allow for this occurrence, and so we must be vigilant, prepared, and looking up from within our hearts to the heavens always, and especially from easter 2026 on. It will be a very intense and interesting time to be alive. The Lady has guided Mr. Bledsoe to this location and timing for a particular purpose that is yet to be revealed. There should be no place that one would rather be than right here, right now.

Key Points of Disclosure:

Dr. Steven Greer's "Asymmetric Disclosure": This is an approach to revealing suppressed information about extraterrestrial life and advanced technologies, which Greer believes have been hidden by governments and covert programs. He argues that this suppression is to protect economic and geopolitical systems reliant on fossil fuels.

Bob Lazar's Claims: Bob Lazar is a well-known *"whistleblower"* who claimed to have worked on a secret government project at a site called S-4, near Area 51, to reverse-engineer alien technology. He alleges that the propulsion systems of the flying saucers he saw were powered by a stable element known as Element 115. While Lazar's claims brought significant public attention to Area 51 and UFOs, he possesses classified information that was later confirmed by the US government. Lazar's claim about Element 115 is often cited as compelling evidence, as the element known as Moscovium was officially synthesized in 2003, years after his initial claim in 1989.

Chris Bledsoe's Prophecy: Chris Bledsoe claims a figure he calls *"The Lady"* informed him of an astronomical event that will occur around Easter in 2026. According to Bledsoe, the star Regulus will appear red on the horizon in front of the Sphinx, signifying *"the end of an old way and the beginning of a new."* Astronomically, Regulus is a blue-white star, but atmospheric conditions could potentially make it appear red. Chapter 19 notes that while the specific prophecy is speculative, celestial alignments are a real phenomenon observed today and by ancient Egyptians. Chapter 19 discussed various aspects of the UFO/ADA and extraterrestrial intelligence disclosure movement, focusing on three main concepts:

- Dr. Steven Greer's *"asymmetric disclosure"*.
- The story of Bob Lazar.
- Chris Bledsoe's encounters and prophetic visions from *"The Lady."*

Chapter 19 considers the Bledsoe prophecy in alignment with predictions from Edgar Cayce regarding a *"Hall of Records"* beneath the Sphinx that would be revealed when humanity reaches a higher level of spiritual consciousness. Greer advocates transparency and the public release of information, believing it would unite humanity and lead to a new era of technological advancement. Lazar claims to have worked on a secret government project at a site called S-4, near Area 51, to reverse-engineer alien technology. When considered together, there is an extraordinarily strong correlation between all these varied ideas that can no longer be simply ignored.

Chapter 20
The Seers - Cycles Connection

Throughout history, seers, mystics, and religious visionaries have delivered warnings of devastating global events yet to come. These prophecies are often interpreted through the lens of modern warfare, with many assuming such destruction refers to nuclear conflict or other man-made catastrophe. However, another interpretation emerges when these messages are viewed alongside the scientific and spiritual framework established in this work: that the devastation described may instead point to cyclical natural cataclysms tied to the Earth, the Sun, the Milky Way Galaxy, and the electromagnetic current sheets that connect celestial systems. This chapter explores that possibility and the growing connection between prophetic revelation and cosmic cycles.

There are bodies of work that seem to prove that the sun does go Nova, or what some say is Micro-Nova, about every twelve thousand years or so. The current thought is that the solar Nova/pole shift will likely last about three days, with subsequent smaller yet global-scale disastrous events that continue for a much longer period thereafter. With the initial focus upon the three days of the worst of the nova/pole shift, let us first turn to the mystics.

The Catholic mystics whose private revelations are most frequently cited regarding a *"three days of darkness"* are Blessed Anna Maria Taigi and Marie-Julie Jahenny. It is important to note that the Catholic Church has not publicly accepted the validity of these specific prophecies, and belief in them is not a required part of the Catholic faith. These seers help us to develop the common threads that make the tapestry of evidence concerning these disaster cycles.

Blessed Anna Maria Taigi (1768-1837): Anna Maria was an Italian mystic beatified by the Church, meaning her life was found to be one of heroic virtue. While she still awaits sainthood, the accounts of her life from her spiritual director that were recorded after her death describe a vision that she had of an intense darkness. The darkness will go over the whole Earth during which only blessed candles could provide light, and the enemies of the Church would perish.

Marie-Julie Jahenny (1850-1941): Marie-Julie is known as the *"Breton Stigmatist."* She reportedly provided more detailed messages from Our Lord about the three days of darkness. Her messages specified that only pure beeswax candles would light and that people should stay inside their homes and pray the rosary for the entire three days. *"The Earth will be covered in darkness... and Hell will be loosed on Earth..." And "Thunder and lightning will cause those who have no faith or trust in My Power to die of fear."* She added more with: *"During these three days of terrifying darkness, no windows must be opened, because no one will be able to see the Earth and the terrible color it will have in those days of punishment without dying at once... The sky will be on fire, the Earth will split... let the blessed candle be lit everywhere, no other light will shine..."* and she goes on to explain that *"No one outside of a shelter... will survive. The Earth will shake as at the judgment, and fear will be great... Yes, we will listen to the prayers of your friends; not one will perish. We will need them to publish the glory of the Cross..."* Finally, she adds that, *"During this darkness the devils and the wicked will take on the most hideous shapes... red clouds, like blood, will move across the sky. The crash of the thunder will shake the earth, and sinister lightning will streak the heavens out of season. The Earth will be shaken to its foundations. The sea will rise; its roaring waves will spread over the continent...,"* and *"The Earth will become like a vast cemetery. The bodies of the wicked and the just will cover the ground. Three-quarters of the population of the globe will disappear."*

Venerable Elizabeth Canori-Mora (1774-1825): Elizabeth is another Italian mystic. Her writings were examined during her cause for canonization and found to contain nothing against faith or morals. Her visions also mentioned great chastisement followed by a restoration of the Church.

Saint Gaspar del Bufalo (1786–1837): Saint Gaspar is also occasionally listed among those who foretold such a period of darkness. There is not much else to be found, and it is strongly recommended that academics pull his records from the Vatican to see what else there may be concerning his accounts in the archives of the Vatican.

Saint Sister Mary of Jesus Crucified (1846–1878): Mary of Jesus Crucified is also known as Mariam Baouardy. She is another figure sometimes mentioned in connection with messages concerning the three days of darkness. There is not much else to be found on this case as well, and it is strongly recommended that academics pull these records from the Vatican to see what else there may be of concern about her accounts as well.

Luz de Maria de Bonilla: Luz is Costa Rican-born and currently lives in Argentina. Her exact age is not publicly disclosed; however, she is a living adult who has been receiving messages since at least 1991. Her messages, which are considered private revelations. Her messages describe the three days of darkness as a time of divine purification and judgment by fire and wind, during which humanity will perish in darkness because of sin. The darkness will last for three days and three nights. She claims that a fire will fall from the sky and rise from the earth, consuming everything and all of man's works.

The air will be laden with pestilence and demons. For her, the event is described as a *"terrible day"* brought about by humanity's stubbornness, sin, and rejection of God's laws. This idea is like that of those who built the Tower of Babel. It is a result of Divine Justice for a *"depraved, licentious, dehumanized, arrogant"* generation. Luz explains that the faithful should remain in their homes, pray the rosary, and beg God for mercy.

The messages emphasize that God is full of mercy and that the punishment can be mitigated if hearts are full of humility and repentance. Luz de Maria's messages fit within a broader tradition of mystics like Blessed Anna Maria Taigi and Marie-Julie Jahenny, who have also prophesied about a period of intense darkness and purification. The Catholic Church offers guidance on private revelations but does not require belief in the three days of darkness prophecy.

Padre Pio: While his prophetic words are mostly a hoax, they are significant in their details for the purposes of connecting these messages to the larger theme of solar and galactic cycles. It is claimed that Parde Pio stated in part, *"Hurricanes of fire will pour forth from the clouds and spread over the entire earth! Storms, bad weather, thunderbolts, and earthquakes will cover the Earth for two days."*

He later is said to continue with: *"An uninterrupted rain of fire will take place! It will begin during a very cold night...."* and *"That you may be prepared for these visitations, I will give you the following signs and instructions."* ... *"The night will be very cold. The wind will roar... thunderbolts will be heard. Lock all the doors and windows. Talk to no one outside the house."* It is also attributed to Padre Pio that he claimed that *"The wind will carry with it poisonous gases which will be diffused over the entire earth. Those who suffer and die innocently will be martyrs; they will be with Me in My Kingdom. Satan will seem to triumph! After three nights, the earthquake and fire will cease...*

The sun will shine again. Angels will descend from Heaven and will spread the spirit of peace over the earth. A feeling of immeasurable gratitude will take possession of those who survive this terrible ordeal..."

The Church emphasizes living a life of faith, receiving the sacraments, and seeking God's will as the best way to be spiritually prepared for the future, rather than focusing on specific apocalyptic predictions from private sources.

Some of What the Bible Says:

Amos 8:9: "In that day," declares the Sovereign Lord, *"I will make the sun go down at noon and darken the Earth in broad daylight."*

Joel 2:2: "...a day of darkness and gloom, a day of clouds and blackness. Like dawn spreading across the mountains, a large and mighty army comes, such as never was in ancient times..."

Jeremiah 13:16: "Give glory to the Lord your God before he brings the darkness, before your feet stumble on the darkening hills. You hope for light, but he will turn it to utter darkness and change it to deep gloom."

The Book of Revelation contains several passages that have been interpreted by some as referring to a period of darkness associated with divine judgment.

Revelation 6:12: Describes a great earthquake, stating, *"the sun became black as sackcloth of hair, and the moon became as blood."*

Revelation 8:12 States, *"And the fourth angel sounded, and a third part of the sun was smitten, and a third part of the moon, and a third part of the stars; so, as the third part of them was darkened, and the day shone not for a third part of it, and the night likewise."*

These verses are often cited in discussions about apocalyptic darkness, though the Catholic Church typically emphasizes spiritual preparedness over literal interpretation of such private revelations. While Revelation contains vivid imagery of cosmic disturbances and darkness, it does not specify a prophecy of *"three days of darkness"* as described in more recent private revelations or traditions. Instead, its language points to symbolic and dramatic events as signs of the end of time, urging believers to remain steadfast in faith and hope. It is the sun cycles and galactic cycles that determine the events that are prophetically described because they are based upon electro-magnetic sheets that are commonly called current sheets.

These magnetic current sheets emanate from celestial objects like the center of galaxies, from stars, and to a lesser degree from planets. These electromagnetic sheets emanate from all celestial objects, both those with an accompanying star or solar system and those that are alone. These cycles have been demonstrated by several researchers, including Doug Vogt, Randall Carlson, Mathew LaCroix, Ben Davidson, Robert Schoch, and August Dunning, just to name a few. Chapter 20 describes connections between prophetic messages from various Catholic seers and theories about cyclical, cataclysmic events caused by the sun and galaxy.

Key Connections:

Prophecy and Solar Theory: The text connects the *"great devastation"* mentioned in prophecies like those from the Virgin of Fatima, and the Lady of Chris Bledsoe, to the idea that the sun undergoes a *"Nova"* or *"Micro-Nova"* event every twelve thousand years or so.

"Three Days of Darkness": Specific Catholic mystics like Blessed Anna Maria Taigi and Marie-Julie Jahenny are cited for their private revelations about a future *"three days of darkness"*. This aligns with the theory that the solar event and subsequent pole shift would last about three days.

Physical Explanations: Chapter 20 posits that the seers are referencing actual physical phenomena related to the Earth, the sun, their interplay with the Milky Way Galaxy, and electromagnetic current sheets.

Biblical References: Passages from the Bible books like Amos, Joel, Jeremiah, and Revelation that mention the sun darkening and cosmic disturbances are included, suggesting scriptural support for the idea of a dark period.

Scientific Researchers: The text names several researchers, like Doug Vogt, Randall Carlson, Ben Davidson, among others, whose works demonstrate these cycles and the related existential threats, linking their scientific theories to ancient and modern prophecies.

Survival Imperative: The ultimate connection is the need for people to prepare spiritually, physically, and emotionally to survive this coming *"end of the age"*. Chapter 20 concludes that the prophecies discussed describe a *"three days of darkness"* event potentially linked to solar cycles, galactic time scales, and pole shifts.

Chapter 21
Can Anything Be Done?

If the cycles and warnings examined throughout this work are accurate, then the question naturally becomes not whether preparation is necessary, but how one should prepare. While this book does not claim to provide definitive expert instruction for surviving extinction-level events, it does seek to present practical and spiritual considerations for those contemplating what may lie ahead. Preparation may take many forms, from physical readiness and logistical planning to spiritual discipline and inner resilience. As humanity approaches what some believe to be the closure of the Great Year through galactic, solar, and terrestrial cycles, everyone must consider which preparations will best position them for what may come.

Key Revelations and Safe Places:

The refuges described by Marie-Julie Jahenny serve as a spiritual and physical blueprint for surviving the prophesied *"Three Days of Darkness,"* which is a warning from a seer of the impending solar crisis.

Appalachian Mountains: These remote valleys and hidden communities in West Virginia, Eastern Kentucky, and Western Virginia will serve as the first refuge. Believers there have been prepared to live simply and depend on God, with supernatural provision of resources like food and water.

Ozark Mountains: Vast, natural underground cave systems in Arkansas and Southern Missouri are the second refuge. These hidden caverns will provide safety when life becomes difficult, with divine provision of fresh air, water, and food, and angels visibly moving among the inhabitants.

Humble Churches in the Heartland: Small, faithful congregations across the American heartland will become fortresses of God's presence, offering supernatural protection against the events to come and providing miraculous healing and provision for large gatherings.

Individual Homes in the Great Plains: Ordinary homes of devoted believers in states like Nebraska, Kansas, Oklahoma, and Texas will be transformed into divine sanctuaries, acting as beacons of light, healing, and evangelism, miraculously expanding to accommodate many and never running short of provisions.

Phases of Civil Unrest:

Phase 1: Technological Collapse: A massive solar storm could cripple the nation's electrical grid and financial systems, leading to widespread chaos and desperation. This is presented as the *"enemy's opening move."* The prepared refuges will remain unaffected, with their inhabitants living independently of failing modern systems.

Phase 2: Religious Persecution: Approximately two months after the initial collapse, emergency legislation will restrict religious gatherings. Technology will become a comprehensive surveillance and control system. The faithful in the refuges will be divinely shielded from this intrusion.

Phase 3: False Religious System: About four months after the collapse, a deceptive movement promoting false unity and "all spiritual paths" will emerge, tempting many believers. The people in the true refuges will distinguish this from genuine faith through their deep understanding of God's word.

Phase 4: Intense Persecution: Approximately six months after the initial collapse, the persecution of genuine Christians will reach extreme levels, leading to imprisonment, torture, and martyrdom. Despite this, the four refuges will continue to shine as beacons of hope, self-sufficiency, angelic visitations, and miraculous power.

Preparation for Refuge: Jesus emphasized that not everyone who claims faith will find safety in these places.

Only those with truly surrendered hearts, tested and proven faith, and a love for Him that exceeds their attachment to worldly comfort and security will be supernaturally guided to these sanctuaries. Sister Marie highlights examples of humble believers who were guided by divine prompts due to their deep faith and simple living. She is contrasting them with those whose attachment to earthly possessions or bitterness prevented them from receiving guidance. The protection within these refuges will intensify, with military satellites malfunctioning, GPS systems failing, and buildings appearing invisible to hostile forces. The refuges will also experience miraculous provision, violating laws of physics and economics. Each type of refuge is assigned to a specific divine role, such as intercession, preservation of truth, healing, and evangelism. While the vision of sister Marie is more concerned with the Anti-Christ, what if the final revelations of the bible are about the cycles of the sun and what will happen? In this light, these refuges serve a much greater purpose for the survival of humanity. According to her revelations, these safe havens are primarily the homes of the faithful that are secured not by modern technology, but by sacramentals and prayer. Ultimately, Jahenny's prophecies conclude that the only true *"refuge"* is a state of grace. The physical preparations are secondary to the divine protection promised to those who maintain their faith, suggesting that the interior life of the believer is the ultimate sanctuary against the impending chastisement or solar nova/pole shift.

Chapter 22
The Two Timelines

When the full body of evidence presented throughout this book is considered together, a larger pattern begins to emerge, suggesting that humanity may be approaching a significant point of divergence. Numerous seers, mystics, and spiritual traditions describe humanity as standing between two possible futures, often portrayed through differing symbolic frameworks. What some call ascension and others describe as biblical end-times prophecy may, in essence, reflect the same transformative process viewed through distinct cultural, religious, and historical lenses. This chapter explores that possibility and examines how these seemingly separate narratives may point toward a unified truth.

Humanity's Two Timelines:

Higher 3D Ascension: Also called the ascension timeline, this timeline is where the 3D Earth progresses towards a more enlightened state, merging with higher dimensions. It represents spiritual growth and awakening, or what the faithful call the rapture and second coming of Christ.

Characteristics: This path is rooted in love, truth, freedom, unity, and progress. It represents a collective shift towards higher consciousness and spiritual awakening.

Personal Growth: Individuals on this timeline focus on personal resonance, aligning their thoughts, emotions, and intentions with positive vibrations. This alignment fosters opportunities for mass awakening and collective evolution.

Collective Impact: The ascension timeline emphasizes the interconnectedness of personal and collective experiences. As individuals raise their frequencies, they contribute to a broader shift in humanity's consciousness.

The next image represents the two timelines and their divergence.

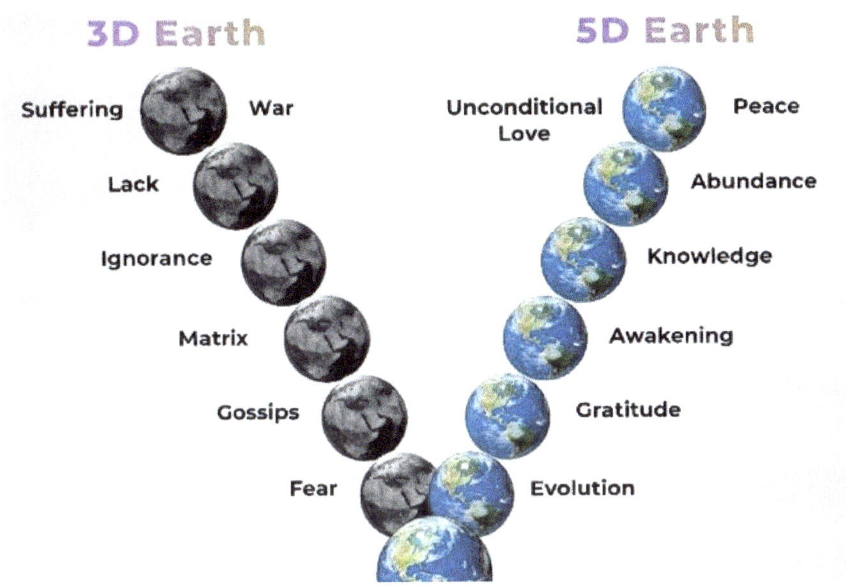

Enhanced image courtesy of: https://timefordisclosure.com/what-isthe-new-earth-5d-delores-canon/

Lower 3D Continuation: Also known as the negativity timeline. The Lower 3D timeline is where humanity remains stagnant and descends into a less harmonious state. This path reflects a lack of spiritual development.

Characteristics: In contrast to the ascension timeline, this path is steeped in fear, illusion, control, division, and stagnation. It reflects a reality where individuals remain trapped in outdated belief systems and external dependencies that become a downward spiral that feeds negative interdimensional entities.

These entities are called demons and demonic influences for the faithful. These evil entities gain power and strength that they use to keep humanity trapped. The faithful call these interactions sin, capital sins, and degrading actions.

Personal Experience: Those aligned with the negative timeline may experience a sense of separation, conflict, and often feel overwhelmed by negative emotions, fear, and societal pressures. The faithful call this demonic oppression and sometimes even possession by evil spirits. The experience of hell is complete and utter separation from God, the Creator of the universe.

Collective Consciousness: There are several factors that influence both timelines. One such influencing force is called the collective consciousness. The ascension of humanity depends on the percentage of humanity that holds higher consciousness. A threshold of 22% is commonly identified as necessary for meaningful change. If humanity collectively chooses to elevate its consciousness, it can shift towards the higher 5D timeline. The faithful call this conversion and turning one's heart toward a personal relationship with the Savior, Jesus Christ. They also believe that God loves everyone and wishes to save all of humanity for eternity. God loves us so much that he allows us to choose our own path, one of life or one of death. This is to say one of ascension or one of devolution.

It is this lack of awareness, and increasing fear, hatred, and war that is used as a means of control. The faithful call this the tribulation and end times that, according to some bible scholars, must happen for Christ to return to judge the living and the dead. The rapture is primarily described in 1 Thessalonians 4:16-17, where it states that the Lord will descend from heaven, and the dead in Christ will rise first, followed by living believers being caught up to meet Him in the air.

Other relevant passages include 1 Corinthians 15:51-52, which speaks of believers being changed in an instant, and Matthew 24:30-31, where Jesus mentions gathering His elect at His return. The nature of the rapture: The term *"rapture"* itself does not appear in the Bible. It is derived from the Latin word *"rapturo,"* translating the Greek word *"harpazo,"* meaning *"to seize"* or *"to snatch away."*

The rapture is often associated with the second coming of Christ but is considered a distinct event where believers are taken up to be with Him. The concept of the rapture in Christianity, particularly in American evangelicalism, originated in the 1830s with the teachings of John Nelson Darby, who introduced it as part of Dispensationalism. This doctrine suggests that believers will be taken up to meet Christ before a period of tribulation, contrasting with earlier Christian views that did not emphasize a pre-tribulation rapture.

Timing and Interpretations: The timing of the rapture is debated among Christians. Some believe it will occur before a period of tribulation; this is a pre-tribulational view, while others suggest it will happen during or after tribulation.

Different Denominations: There are various interpretations of these passages in many ways, with many mainstream Christian groups not subscribing to the rapture as commonly understood in evangelical circles. The rapture is a significant eschatological event in Christian theology, emphasizing the hope of resurrection and eternal life for believers. The key biblical passages provide a foundation for understanding this doctrine, though interpretations vary widely among different Christian traditions.

Individual Responsibility: The actions of each person and their thoughts contribute to the overall timeline. Positive behaviors based on love and spiritual growth can help steer humanity towards ascension. Conversely, negative actions and a lack of awareness can reinforce the lower 3D timeline. This, concisely, is another way to express free will. We all make a conscious choice in every action, every day, for love or for hatred. The choice is for everyone to make, and by not choosing, you will be making a choice, even if you do not believe it.

Collective Consequences: The negativity timeline leads to a collective entrenchment within fear-based systems, hindering progress and severely slowing, if not entirely stopping, spiritual growth.

The devolutionary path can create barriers to awakening and limit the potential for a unified, enlightened existence. Personal choices play a crucial role in determining which timeline you align with. As an individual, your free will, your choices shape your frequency, beliefs, emotional state, and personal relationship with the Creator of the universe. The faithful call this being lost in the world or being found in the love of God. By making choices that resonate with love, truth, and higher consciousness, you can navigate towards a more positive ascension timeline. Choices that are rooted in fear and negativity will keep you anchored in less favorable realities, in a lower frequency, and susceptible to fear, hate, and the controlling influence of external evil forces. Potential future scenarios for humanity include risks from artificial intelligence, environmental degradation, and global catastrophic events like nuclear war, pandemics, and solar events. These scenarios will lead to significant societal disruption, economic collapse, or even human extinction if not addressed properly.

The choice between the ascension and stagnation timelines is not fixed. The movement to a particular timeline is shaped by collective human consciousness and individual choices that impact the collective towards love or towards evil. The potential for a brighter future exists if humanity collectively strives for higher awareness and spiritual growth. The ascension timeline for human evolution involves a gradual increase in frequency and consciousness, with significant shifts occurring over the past few decades. The faithful call this spiritual revival. This process is marked by collective awakenings and the transition from lower to higher frequency bands, ultimately leading to a more harmonious existence. Many Marian apparitions contain this very message of conversion, prayer, and turning away from evil. Choosing a negative timeline for human devolution will lead to catastrophic events such as solar novae and pole shifts that will likely result in the loss of biodiversity and severe climate changes that threaten all life forms on the earth. This idea is corroborated by Chris Bledsoe and the messages he has received from the Lady concerning the two timelines as well.

Essentially, all the previous sections of this book are the who, what, where, and how of stagnation on the human devolution timeline. There are also impacts that must be considered from the control of the narrative on the ascension timeline. The limiting factors of the bible may hinder personal growth and ascension to higher states of consciousness for some. By controlling the narrative from a biblical perspective, there are limits set in the understanding of spiritual experiences and the broader interpretations of faith that can then also limit relations with the creator of the universe, which in turn will slow spiritual growth and the ascension process. A strict biblical frame of reference can create a rigid framework that does not allow for the exploration of diverse spiritual paths and insights.

The Governmental Connection:

The U.S. government officially utilized remote viewing through the Stargate Project and related programs like GRILL FLAME and SUN STREAK from the 1970s until 1995. While the official goal was intelligence gathering, researchers and former participants often discuss more complex temporal concepts, including *"precognitive remote viewing"* and theories regarding multiple timelines. In the context of government-related remote viewing, the concept of *"two timelines"* usually stems from Project Looking Glass, which is a rumored, non-officially declassified offshoot, and the declassified Gateway Process documents. Whistleblowers and proponents of the *"Looking Glass"* theory claim that technology was used to see multiple possible futures. The *"two timelines"* often refers to a theory that multiple paths existed until a *"convergence point,"* which is sometimes cited as 2012 or 1999, after which reality allegedly settled into a single, potentially *"artificial"* path.

Project Looking Glass is an alternative theory centered on an alleged secret military program capable of viewing future events and alternate timelines. The narrative primarily stems from the testimony of self-proclaimed whistleblower Dan Burisch, also known as Dan Crain, who claimed to have worked on the project within classified facilities such as S-4 near Area 51.

Core Technology and Claims:

The Device: Allegedly a physical apparatus consisting of a large, dark gray metal circular ring with a core of spinning electromagnetic rings. Inside, a rotating barrel lined with quartz crystals and argon gas purportedly created a *"spacetime distortion"*.

Capabilities: The system reportedly functioned as a *"time-viewer"* or *"timeline simulator,"* projecting visual images of probable future outcomes based on current global decisions.

Origins: Claims suggest the technology was reverse engineered from recovered extraterrestrial artifacts, specifically ancient Sumerian cylinder seals that contained instructions for accessing *"natural stargates"* or wormholes.

The Two Timelines:

The theory emphasizes a critical divergence in human history often referred to as the *"Two Timelines"* or the *"2012 Convergence"*.

Timeline 1 (The Catastrophic Path): This timeline allegedly foresaw a global cataclysm, often described as a massive solar event or pole shift occurring around December 2012, which would lead to the destruction of civilization or the division of humanity into different evolutionary paths.

Timeline 2 (The Awakening/Averted Path): Project operators supposedly discovered that as history approached December 2012, all probable futures began to converge into a single, unavoidable outcome regardless of elite manipulation. This convergence reportedly signaled a *"Great Awakening"* or a shift in human consciousness that prevented the predicted catastrophe, leading to our current reality, where the predicted apocalypse never occurred.

Control Defined:

Limiting Spiritual Exploration: A rigid narrative may discourage individuals from exploring diverse spiritual experiences. This can hinder personal growth and understanding of a higher dimension.

Suppressing Alternative Views: There are many missed opportunities where valuable insights from other spiritual traditions are overlooked. Because of the prioritizing of a singular interpretation of faith, they are entirely ignored or flat-out rejected as heresy. This can create a narrow view of spirituality that does not align with the expansive nature of a higher dimension.

Fear of Divergence: A focus on maintaining control can instill fear in believers about questioning or deviating from established doctrines. This fear can prevent individuals from pursuing their spiritual journeys authentically. There can be consequences for spiritual growth because of a controlled narrative that can manifest in many ways:

Stagnation in Faith: When individuals feel confined by dogma, their spiritual growth may stagnate. This can lead to a lack of engagement with deeper spiritual truths that lead to the lower timeline referred to earlier.

Disconnection from Community: A controlled narrative that excludes differing perspectives can alienate members of the faith community. This disconnection can hinder collective spiritual ascension.

Resistance to Change: There is a reluctance among the faithful to adapt to a change in narrative in response to contemporary spiritual needs, creating resistance, jealousy, and animosity among believers seeking a more profound connection with the divine.

Key Takeaways:

Controlling the narrative from a biblical perspective can negatively impact the ascension to a higher dimension by limiting spiritual exploration, suppressing alternative views, and fostering fear, ultimately hindering personal and communal spiritual growth. This concept suggests that such tribulations serve as a purification process, preparing individuals for a deeper connection with the divine and a transition to a higher dimension of existence. Different spiritual beliefs interpret the timelines in various ways. For instance, Young Earth Creationists believe the Earth is around 6,000 years old based on a literal interpretation of the Bible, while Old Earth Creationists and Theistic Evolutionists accept scientific evidence suggesting the Earth is approximately fourteen billion years old. The old earthers view the biblical creation account in Genesis as compatible with this understanding. Humanity is currently experiencing a divergence between these two timelines. Love is a focus on ascension, characterized by the eternal or divine, unity, and higher consciousness. Devolution is rooted in negativity, fear, hate, and control. This split reflects the collective and personal choices that individuals make, influencing the reality they experience and the future of humanity, where some will remain in the lower timeline while others rise to the higher timeline. The faithful sometimes refer to this as the last judgment, where the goats will be separated from the lambs.

These references are both about the two timelines and how some will be on one timeline, and others will be on the other timeline, based upon individual choices and spiritual growth.

Chapter 23
Comparing Ideologies

There are moments in history when people begin to sense that something larger is unfolding beneath the surface of everyday life. Across different traditions, certain patterns begin to repeat, especially through the voices of seers who describe events that point toward both warning and transformation. Within the Catholic faith, two of the most widely discussed messages are the Three Days of Darkness and the Illumination of Conscience. These are often understood as events that lead into a deeper spiritual awakening, where individuals are brought face to face with their relationship to the Creator and called toward repentance, reflection, and change.

Correlation with Spiritual Awakening		
Aspect	**Three Days of Darkness**	**Illumination of Conscience**
Nature of Event	A period of darkness and tribulation	A moment of clarity and self-awareness
Purpose	**Purification and judgment**	**Invitation to repentance and spiritual growth**
Outcome	Spiritual renewal for the faithful	Awakening to one's true spiritual condition

Comparative Consideration:

Both events are seen as part of a larger spiritual journey. The darkness serves as a catalyst for introspection and repentance, while the illumination offers clarity and understanding. Together, they are believed to facilitate a transition to a higher state of consciousness or spiritual awakening. It is this awakening that is aligning with concepts of ascension to higher dimensions. The Three Days of Darkness and the Illumination of Conscience are interconnected prophetic messages that emphasize the need for spiritual readiness and transformation.

A spiritual awakening is ultimately the guiding principle for individuals to begin the journey to a higher dimension. Practices that can enhance the experience of ascension include daily meditation to connect with your inner self, engaging in deep breathing exercises, and maintaining a healthy lifestyle with a focus on organic foods and hydration. Additionally, exploring spiritual gazing and primal rhythms, like drumming, can help expand consciousness and facilitate spiritual experiences. Spiritual awakening and ascension, as understood in many New Age contexts, may not align with Catholic teachings, which emphasize a relationship with God through Christ and the Church rather than personal spiritual evolution. Catholicism focuses on salvation and spiritual growth within the framework of faith, sacraments, and community rather than individual ascension experiences. Spiritual awakening and ascension are concepts often associated with personal growth and enlightenment. In many spiritual traditions, these terms refer to a heightened awareness of one's connection to the divine and a transformation of consciousness. However, their compatibility with Catholicism can be complex. A Catholic perspective on spiritual growth emphasizes a journey that is rooted in a relationship with God through Jesus Christ. This journey involves the following:

Faith Development: Catholics believe in the importance of faith, prayer, and the sacraments as means to grow closer to God.

Transformation: Spiritual growth is seen as a gradual process of becoming more Christ-like, which includes moral and ethical development.

Community: The Church community plays a vital role in supporting individual spiritual journeys.

Ultimately, while the Three Days of Darkness and the Illumination of Conscience serve as powerful catalysts for transformation, the path one takes following these events depends on one's spiritual framework. For those in modern spirituality, these moments function as a bridge toward personal ascension and a higher dimension of consciousness achieved through individual practices like meditation and primal rhythms. In contrast, the Catholic perspective views this awakening not as a solitary evolution but as a call to deepen one's relationship with God through Christ. Whether framed as an individual shift in vibration or a communal journey of faith and repentance, both paths underscore a critical need for spiritual readiness in the face of a changing world. Both paths are the same means to an end, eternal life.

Chapter 24
Return to the Divine

In recent discussions, the understanding of ADAs has begun to move beyond purely technological or extraterrestrial explanations into something far more complex. What was once viewed through a scientific lens alone is now being reconsidered with spiritual implications, raising questions about faith and the deeper forces at work in the world. There is a growing sense that fear surrounding these phenomena may be shaping public perception in ways that serve control rather than truth. This shift invites a unique way of seeing, where the focus moves away from fear and toward a path rooted in faith, awareness, and a return to a more unified understanding of the divine.

There are what some might label as attempts at a false flag warning in the name of defense and the protection of the Earth, as well as the greater solar system. A fear-based approach is a blatant lie that is propagated to exercise control that must be disputed and refuted. Fear must be extinguished if we are to become a people of love and peace on a global scale. There are two ways to see the world and to live. One is the way of fear, control, hate, and war. The other is the way of love, peace, and unification that is respectful of nature and the divine found in all of creation. One can tell by the frame of reference of others, where they live, by fear or by love. Fear not, for I go before you, says the Lord, and maybe these ADAs are simply God going before us to make straight the paths for us? When the message is conversion, prayer, fasting, and a return to love, is there any other way to consider these divine calls that are based upon authentic love?

Spiritual Elucidations:

Some researchers, like Diana Pasulka, a professor of religious studies, have explored the idea that ADA phenomena may have a spiritual dimension, drawing parallels between contemporary accounts and historical religious texts or beliefs. Like her works is the interdimensional hypothesis. The interdimensional hypothesis suggests that ADAs are manifestations of beings or phenomena from other dimensions that coexist alongside our own. This is a concept that resonates with some spiritual and esoteric traditions.

It is important to note that discussions about the spiritual aspects of ADAs often involve interpretations and may not align with mainstream scientific or religious views. This misalignment is usually because of tradition or scientific proof that would move a hypothesis to mainstream theory that does not yet exist. For the believer, this information should be a means to expand faith and to minimize separation and doubt. Both mystical experiences and ADA encounters share some common ground in their potential to induce altered states of consciousness. Altered states often lead to encounters with the supernatural and foster a search for meaning and transcendence. However, they remain subjects of ongoing exploration and debate, with diverse interpretations and a need for further research.

Supernatural Experiences and Appearances:

There also exist other commonalities shared by the apparitions of Chris Bledsoe, Medjugorje, Fatima, and Lourdes, and more contemporary visionaries that must be considered in the totality of current events. One must acknowledge the distinct characteristics of each. There are levels of ecclesiastical approval that, for the Catholic Church, are those encounters that are officially approved by the Church and those that are called private revelations that can be considered by the faithful but are not formally approved by the church. They all have some framework that connects them because nearly all these accounts involve individuals claiming to have experienced contact with beings or phenomena beyond ordinary perception. In Lourdes, Fatima, and Medjugorje, the visions are primarily of the Virgin Mary, often with specific messages and instructions. Chris Bledsoe's experiences, while initially focusing on ADAs and orbs, also include encounters with a female figure he calls *"The Lady,"* raising questions of potential Marian or angelic connections. Other mystic visionaries are visited by the Virgin Mary, Saint Michael, and at times by Jesus Christ with varying degrees of warnings, instructions, and calls to return to love.

Spiritual Transformation and Messages:

A central theme across all these events is a call for spiritual change, including prayer, conversion, and a deeper connection to the divine.

The Marian apparitions at Fatima, Lourdes, and Medjugorje emphasize prayer, especially praying the Rosary, penance, and seeking peace. Chris Bledsoe's encounters, particularly with *"The Lady,"* also have a spiritual dimension, leading him to call all who will listen to reflect on themes of love, compassion, nature, and the universe. Other mystics report messages of impending punishments for the lack of faith of all of humanity. Some have suggested that these punishments are a prophecy of coming wars and end times. A deeper look into these messages suggests something far more specific than war that is impending on the Earth. Judgment or not, there are certain cyclical events that must be taken into consideration when reflecting upon these messages. In the totality of this book, these messages are warnings of both the sun, the cycles, and the sciences involved in what these seers call to mind, which is that the Sun does Nova. Some call it Micro-Nova, and terms are relative at this point because there is no real consensus on how stars really work.

Impact on Witnesses and Communities:

The individuals experiencing these apparitions often undergo significant personal transformations and may face scrutiny, disbelief, and even ostracism from their communities or religious institutions. Despite potential challenges, these experiences frequently lead to increased faith and spiritual growth for the visionaries and those who believe their accounts to be true and divine. Lourdes and Fatima have become pilgrimage sites and hold significant religious value for many Catholics. Medjugorje, while not yet formally approved by the Catholic Church as supernatural, also attracts many pilgrims who report spiritual experiences and conversions. Chris Bledsoe has also been contacted by several government agencies concerning his experiences with *"The Lady"* and how her messages may impact us all. There may be some connections to prophetic words and things like remote viewing that need to be explored as connected in greater detail.

Potential for Healings and Miraculous Events:

Healings are associated with Lourdes, famously linked to the spring water the Lady instructed Bernadette to dig up. Chris Bledsoe's initial encounters include healing of an illness, and he has reported being able to heal others subsequently.

A major miracle is the dancing of the Sun at Fatima. There are suggestions that what the public saw at Fatima was the Merkabah or Cherubim craft that the *"Lady"* arrived in, and its intense heat, light, and glow caused many to think that it was the Sun. While not the primary focus of the messages, miraculous events or healings are reported by some pilgrims to Medjugorje. According to the United States Conference of Catholic Bishops, the Vatican sees spiritual value in Medjugorje, even though it does not judge it as supernatural. While the nature of the encounters and the specifics of the messages may differ, the experiences of Chris Bledsoe,

Medjugorje, Fatima, Lourdes, and others share common threads of perceived supernatural phenomena.

Such common threads also include spiritual transformation and significant impact on individuals and communities. It is important to note the varying levels of formal recognition by religious institutions, with Lourdes and Fatima being formally approved Marian apparitions within Catholicism, while Medjugorje remains under review, and Chris Bledsoe's experiences are interpreted within a broader context of spirituality and the supernatural, as he is not Catholic.

The Private Revelations:

Private revelation refers to divine messages or insights given to individuals or groups, distinct from public revelation, which is intended for all humanity. In Christian theology, particularly within Catholicism, private revelations are seen as personal communications from God, the Virgin Mary, saints, or angels. Sometimes the receivers of private revelations are referred to as seers, and their messages can be frightening for some and an inspiration for others. Because the Church does not officially approve these seers, we, the faithful, have the option to accept or not accept the messages from these mystic visionaries. There are several of them that deliver messages on a regular basis. Like the other experiences above, these messages speak of coming judgments, and things like the three days of darkness, or illumination of all conscience. These seers often speak of wars, earthquakes, famine, and several other possible catastrophic events that are about to unfold upon the world. Most often, these messages end with an exhortation to pray, repent, fast, and prepare.

Nearly all have some level of correlation to prophetic events from the book of Revelation and other end-times prophets like Daniel. There are commonalities of these messages combined with the information in the various parts of this book, where we look at the science of Galactic and Solar Cycles and how these prophetic words need to be reconsidered with those cycles as the focal point for the messages received.

Mystical Experiences & ADA Encounters:

Comparing mystical experiences with All Domain Anomalies (ADA) involves examining how people perceive and interpret extraordinary events, often with a spiritual or transcendent dimension. This may require some of us to allow for the space in our understanding for things that may not fit well with a traditionalist approach. Because the sciences do not align well at times with some traditional approaches that may be overly dogmatic in nature, this does not make these ideas mutually exclusive. Taking a more unified approach does not see new findings as wrong or evil, just as new, and different. If the Church wants to remain relevant in the coming decades, it must move to find common ground that continues to bring these various considerations together. Between the previously discussed historical accounts and modern thought, there is a need to see how these things are all related and point all of us towards a deeper relationship with the creator of the universe. In the end, it is all about living in the light of divine love that always vanquishes the lies of darkness, hate, and control in this world. There is an opportunity to transform our reality, and these tools are meant to assist us all in this journey of human ascendance.

Key Concepts from Chapter 24:

ADA - Interdimensional Hypothesis: The interdimensional hypothesis suggests that ADAs are manifestations of beings or phenomena from other dimensions that exist alongside our own. This concept, which has been proposed by ufologists such as Jacques Vallée and J. Allen Hynek and is still considered today, resonates with some spiritual and esoteric traditions. Proponents of this hypothesis argue that ADAs may be a modern version of a phenomenon that has been reported throughout history and was previously attributed to mythological or supernatural beings.

Diana Pasulka's Research: Diana Pasulka, a professor of religious studies, has explored the spiritual aspect of ADAs by drawing parallels between contemporary accounts and historical religious texts. Her work, including the book *"American Cosmic,"* examines how belief in UFOs is a modern form of spirituality and how these encounters are transformative for the people who experience them. Pasulka suggests that historical accounts of angelic encounters by mystics may have been descriptions of UFO-like entities. She has also noted that some Vatican officials interpret ADA sightings through an angelic or demonic lens.

Common Ground with Mystical Experiences: Both mystical experiences and ADA encounters can lead to altered states of consciousness, encounters with the supernatural, and a search for meaning and transcendence. This is a challenge to traditional science and religious beliefs.

Connections to Religious Apparitions:

Shared Themes: Experiences like those of Chris Bledsoe and the apparitions at Medjugorje, Fatima, and Lourdes share commonalities such as contact with beings beyond ordinary perception. They often include a call for spiritual transformation as well. These events often emphasize prayer, conversion, and a deeper connection to the divine.

Physical and Personal Impact: Witnesses to these events often undergo significant personal transformations and may face scrutiny and ostracism from their communities. Miraculous events and healings are also associated with some of these apparitions, such as the spring water at Lourdes or the dancing of the Sun at Fatima. Chris Bledsoe's encounters also include a healing experience and encounters with a female figure he calls "The Lady," raising questions of potential Marian or angelic connections.

In Chapter 24, we discussed how *"All Domain Anomalies"* (ADAs) have expanded to include spiritual, multi-dimensional, and benevolent interactions, moving beyond purely technological or extraterrestrial explanations. This chapter suggests that a fear-based approach to ADAs is a form of control exerted by those in power to manipulate the public.

A more unified approach is needed to see these phenomena as a path toward a deeper relationship with the divine and a transformation of human reality that must be based on love that raises consciousness and brings all people together. The chapter concludes that the Church must find common ground with new scientific findings to remain relevant in the coming decades, rather than viewing them as mutually exclusive. The messages from these encounters often speak of future events, such as wars, catastrophic events, or even a solar nova, but also contain exhortations to prayer, repentance, and preparation.

Chapter 25
Exploring the Similarities

There are times when different experiences begin to mirror each other in ways that are difficult to ignore. What appears separate at first can reveal shared patterns when examined more closely, especially when both spiritual encounters and modern reports point toward something beyond ordinary understanding. Looking at these similarities allows for a deeper exploration of what these experiences may truly represent.

Altered States of Consciousness and Profound Experiences: Both mystical experiences and ADA encounters can involve altered states of consciousness, where individuals feel they have encountered something beyond ordinary reality. These experiences are often reported as having a deep personal significance, leading to a shift in the individual's perspective or worldview. They are often accompanied by a sense of lost time and not remembering much of the occurrence. Some may call this the awakening process, where there is involvement of the Pineal gland and its decalcification. This process allows some experiencers to feel things that the rest of us normally do not perceive.

Encounters with Elusive or Unknown Entities: In both cases, individuals report encountering phenomena or entities that challenge conventional scientific explanations.

This could involve visions, encounters with perceived *"beings of light,"* or feelings of connection to a higher power or cosmic force. Often, these encounters fill the experiencers with peace and love that is beyond human comprehension. It is this completeness and trust that allows for a greater exchange between the experiencer and the visitors. This overwhelming feeling of authentic love is often accompanied by visions, or of being taken to other places. The experiencer is often shown a potential future event or events that serve as a basis for the changes that some experiencers go through, and their desire to share the knowledge gained with the entire world.

Psychological Impact and Transformation: Mystical experiences and ADA encounters can both lead to profound psychological and even spiritual transformations. Some individuals report a sense of spiritual awakening or a renewed sense of purpose after such encounters. There are some messengers who ask the experiencer to go public with specific information as well. One such case is that of Chris Blesdoe. All these transformative processes are an apparent means to prepare the experiencer for the challenges that come with publicly sharing specifics that may be contradictory and even at odds with current religious or scientific beliefs.

Search for Meaning and Transcendence: An encounter with the unknown or the transcendent, whether through mystical experiences or ADA sightings, can motivate individuals to seek deeper meaning in life. Clearly, the historic perspective of purpose is more widely accepted in the religious context of these events when seen as divine or heavenly messages received from those known by the faithful as having special graces and wanting to impart or admonish humanity to remember the eternal nature or transcendence of our existence.

Narrative and Cultural Resonance: The narratives surrounding both mystical experiences and ADA encounters often draw on cultural myths, religious themes, and a sense of wonder. These stories can explain the inexplicable, offering comfort or a framework for understanding the unknown. The desire to understand these phenomena and their implications for human existence can be a driving force behind exploration and belief. There is also the idea that whoever these visitors are, they most often appear as what is most familiar and loved authentically to the experiencer.

It has been reported that this is done to allow the experiencer to be open to the messages that are being transmitted to them. The recognizable tends to give a certain level of safety, comfort, and understanding that helps to make the messages relevant to everyday life at the time that the message is given. There is also an apparent intent to show relevance into the future when a message speaks prophetically about events to come. There are clearly opportunities created when considering all these factors in the totality of the various experiences over time. The bigger question is, how are we as a people moving to greater understanding into the future with this knowledge of how the divine intervenes in our lives for the greater good of all of humanity?

Where to Look for Proofs: First and foremost is the Bible, because when it reads things like, *"And the Lord took me away in the spirit, into the desert,"* this could be an elucidation from being in an altered state of consciousness. Every attempt will be made to include book, chapter, and verses for your consideration in the reference section at the end of this book. The next place to look is where firsthand accounts become more important, and the works of Diane Pasulka, Chris Bledsoe, Jacques Vallée, and J. Allen Hynek are a great starting point to see how this effect is described by some experiencers. Most experiencers come out the other end a vastly different person than they were prior to having these encounters.

Key Points of Chapter 25:

Altered States of Consciousness: Both phenomena can involve altered states of consciousness, where individuals feel they have experienced something beyond normal reality. Experiencers often recall gaining an increased sense of presence that allows them to know when they are near. These experiences are sometimes associated with the pineal gland, which some believe is related to heightened awareness and spiritual experiences.

Encounters with Unknown Entities: Individuals report encounters with phenomena or entities that are not easily explained by conventional science, such as *"beings of light"* or a connection to a higher power. These visitors often appear in a form that is familiar and comforting to the experiencer, allowing them to be more receptive to the messages being transmitted.

Psychological and Spiritual Transformation: Both mystical experiences and UAP encounters can result in profound personal changes, including a sense of spiritual awakening or a renewed purpose in life. These transformative processes are a means to prepare individuals for the challenges that may arise from publicly sharing their experiences.

Chapter 25 discusses how research suggests that people who report such encounters, referred to as *"seers,"* may have a different psychological profile with higher levels of dissociative and supernatural beliefs compared to a control group. Modern seers often search for meaning in their encounters with the unknown. Such experiences can motivate individuals to seek a deeper meaning in life. The narratives surrounding both types of experiences often draw upon cultural myths, religious themes, and a sense of wonder.

Cultural Context: Throughout history, people have interpreted unexplained aerial phenomena through the lens of their cultural and religious frameworks, seeing them as divine beings or heavenly messages.

Chapter 25 points to evidence for such experiences that can be found in religious texts like the Bible, as well as in the firsthand accounts documented by researchers such as Diane Pasulka, Chris Bledsoe, Jacques Vallée, and J. Allen Hynek. Modern UAP narratives, in a more secular age, may be seen as a new form of spirituality that integrates older religious traditions and scriptures with a technological worldview. In Chapter 25, we also discuss that both mystical experiences and encounters with ADAs share several commonalities, often leading to similar psychological and spiritual transformations in the individuals who experience them.

Chapter 26
Connecting with the Divine

Throughout history, people have tried to make sense of experiences that feel too consistent to ignore yet too complex to fully define. When these events are placed next to one another, a pattern begins to take shape that suggests something more unified than previously assumed. Rather than seeing spiritual encounters, visions, and unexplained phenomena as separate categories, there is a growing argument that they may be connected through a deeper structure that reflects an ongoing interaction between humanity and the divine. This perspective invites a shift in how both religious and scientific ideas are understood.

Viewing the historical angelic, modern UAP, visions, and prophetic phenomena through the lens of a divine family that intervenes in human affairs offers a new perspective on understanding the sacred, the extraordinary, and the supernatural. This proposition challenges the conventional compartmentalization of these experiences. Too often, these experiences are categorized as either scientific anomalies, religious dogma, or subjective mysticism. Maybe advocating for a more unified interpretation is what is needed.

This re-evaluation in our frame of reference could potentially lead to a paradigm shift in both religious and scientific understandings, prompting a reconsideration of long-held assumptions about the nature of reality, divine interaction, and human experience in the context of the totality of historical angelic, modern UAP, visions, and prophetic phenomena. If you are interested in exploring a connection with *"The Lady"* or the divine, there are various paths and practices within spirituality that individuals use to cultivate this connection.

The Bledsoe approach: Mr. Bledsoe was initially visited by three similar craft. This Triad is often seen and may represent the divine holy family that we speak of. He even said that at one point, it was as if there were three suns in the sky. The full extent of his experience can be found in his book, *"UFO of God."* He expresses that there are some options one can utilize to be more able to become an experiencer as well.

Practices for Spiritual Connection:

Meditation, Mindfulness, Prayer: Regularly engaging in meditation, mindfulness, and prayer practices can help to quiet the mind. With a quiet mind, one can become more attuned to inner wisdom and intuitive senses. Prayerful spiritual meditation techniques like asking God for peace, focusing on your breath, and letting thoughts pass can deepen this connection. It has been suggested by some that the Hebrew word for God is the sound of your breath. When you breathe in, you pray Yaw, and when you exhale, you pray Whew. This makes every breath a prayer to God by saying the name Yawhew with every cycle of a breath in, followed by a breath out.

Journaling: The practice of writing down your thoughts throughout the day to go back, read, and reflect upon them later. Journaling can be a powerful tool for self-reflection and connecting with your intuition, allowing you to explore your inner world and identify patterns in your thoughts and emotions.

Time in Nature: Spending time in nature can help you feel more grounded and connected to the rhythms of the natural world, which can be deeply aligned with the divine. Because the divine is love and nature is a large part of the created forces of that love, it is a vital connection that leads both to the divine and leads us directly back to the creator of the universe.

Creative Expression: Engaging in creative activities such as art, writing, or dance can allow you to express your inner self and connect with the creative energy often associated with the divine. By finding these creative connections, one becomes naturally closer to the divine source.

Compassion and Self-care: Embracing self-love, self-acceptance, and self-compassion are important aspects of nurturing the divine within. Fostering kindness, empathy, and understanding towards yourself and others is the beginning of authentic brotherly love.

Understanding the Divine: Learning about the qualities associated with the divine across different spiritual traditions can help you to identify and cultivate these attributes in your own life. These qualities often include empathy, intuition, creativity, collaboration, emotional intelligence, and nurturing. A religious discipline is a great starting point to gain understanding.

Practical Considerations:

Discernment: This is knowledge given by supernatural processes that something is real, beautiful, and good. It is important to develop discernment and critical thinking when exploring spiritual experiences. There are risks associated with unguided spiritual exploration that one must mitigate for your own protection. The basic notion here is that if your experience is not based upon love and light, then it is not divine in its essence and should be abandoned for those experiences that are of love and light.

Authenticity: Some call this *"Know Thyself."* The idea is to approach all of life in general with a trueness of who you are. This is directly connected to compassion and self-care. Specifically, authenticity in spiritual practices involve avoiding the use of spirituality to bypass or avoid dealing with emotional issues. If you have anger and hate inside, then there are issues within yourself that need work.

Mentorship and Community: Consider seeking guidance from experienced spiritual mentors or joining supportive communities focused on spiritual growth to help navigate your journey. A good church, or a close group of people seeking the same disciplines, is often of great assistance. It is important to note that connecting with the divine can be a profound and enriching experience. It is also vital to approach a greater connection with a balanced and discerning perspective. Begin by prioritizing your well-being and seeking guidance when needed. Carl Jung, the Swiss psychiatrist and founder of analytical psychology, extensively explored the concept of the divine within his work on archetypes and the collective unconscious.

What Carl Jung Says About the Divine:

Universal Archetype: The divine feminine, according to Jungian psychology, is a universal archetype residing within the collective unconscious of all individuals, regardless of their biological sex. Think of this as two sides of the same coin, with one side being the masculine and the other side being the feminine, with all of us.

Embody Intuition, Creativity, and Nurturing: The divine feminine represents qualities such as intuition, creativity, receptivity, nurturing, compassion, and emotional intelligence. These are all special attributes that can be developed to their full potential with discipline, practice, and time.

Suppressed in Modern Society: Jung observed that these feminine qualities are often undervalued and repressed in patriarchal societies, leading to a societal imbalance. The opposite is also true in these cases where the masculine is overvalued, and masculine activities are rewarded unduly to reinforce the notion that the masculine is superior to the feminine traits within us all.

Importance of Integration for Balance: Integrating the divine feminine alongside the divine masculine is crucial for individual wholeness and societal harmony. The divine masculine embodies qualities like logic, action, and strength. This is not about one being superior to the other, but about recognizing their complementary nature. This idea is comparable to the yin and yang in Chinese spiritual thought.

Connects to the Psyche and Spirituality: Jung saw the divine feminine as intrinsically linked to the human psyche and spiritual experiences, often manifesting in myths, goddesses, and symbolic representations across cultures. These experiences are how balance is sought in creation because we have chosen to be unbalanced with a societal focus on the masculine.

Examples of Feminine Archetypes: Jungian psychology identifies various feminine archetypes, such as the mother, the Maiden, and the Wise Woman, each representing distinct aspects of the divine feminine. In essence, Jung emphasized the importance of recognizing and integrating the divine feminine within oneself and society to achieve psychological and spiritual balance and foster a more harmonious and fulfilling existence.

In Chapter 26, we have learned that viewing the historical angelic and modern UAP phenomena through the lens of a divine family that intervenes in human affairs offers a new perspective on understanding the sacred and the supernatural. Chapter 26 discusses a new perspective that suggests historical angelic phenomena, modern ADAs, visions, and prophetic messages may all be interconnected manifestations of a divine family. This approach challenges traditional compartmentalized understandings of these experiences as more than simply a scientific anomaly, religious dogma, or subjective mysticism. The chapter outlines several paths for individuals to cultivate a spiritual connection with the divine. Furthermore, Chapter 26 references Carl Jung's psychological perspective on the divine, explaining that the divine feminine is a universal archetype in the collective unconscious. Jung's view emphasizes that integrating divine feminine qualities like intuition, creativity, and nurturing alongside the divine masculine of logic, action, and strength is essential for achieving individual wholeness.

It is individual wholeness that will, in turn, bring balance to the greater society. The chapter concludes by reasserting that examining these phenomena through the lens of a divine family reveals patterns that suggest their recurring intervention throughout history, offering a new way to understand the sacred and the supernatural. The following introductory actions are just some of the ways that you can exercise your mind, grow your frame of reference, and deepen your faith. The beginning is to read, then read more, and then read even more. It is natural to experience self-doubt and fear at different points in the journey of awakening; however, there are numerous strategies and approaches to overcome these challenges and build self-confidence and resilience. There are some other options that you may also find to be of assistance, as follows:

Overcoming Doubts and Fears:

Incorporate Mindfulness and Self-care Practice: Engage in relaxation, prayer techniques, practice deep breathing, or other relaxation methods to manage stress and pray that the Holy Spirit relieve stress and calm your mind and body. This may include reading or a religious practice like Catholicism. Regularly engage in mindfulness exercises or meditation to become more present. Observe thoughts and emotions without judgment and reduce stress. This may also include reading or religious practice to enrich and grow your faith overall.

Shift Your Mindset and Thoughts: Cultivate a growth mindset. Embrace the belief that your abilities can be developed and strengthened through effort and dedication. The key is to be disciplined like St. Benedict or other saints with spiritual teachings. The idea is to choose a discipline, and more than learning it, you live it.

Seek Professional Help if Needed: If self-doubt or fear becomes debilitating, consider seeking support from a therapist, a counselor, or a priest who can provide personalized guidance and strategies. Fear is not of the Lord, and the sacraments of the Catholic Church are a terrific way of help as they provide you with the graces needed to overcome.

Focus on Your Strengths and Achievements: Create a list of your accomplishments and positive qualities to remind yourself of your capabilities. You can do all things through Christ who strengthens you.

Visualize Success: Mentally rehearse desired outcomes, picturing yourself successfully achieving goals to boost confidence and performance. This is where your mind has the power to shape reality and is a major contributing factor to what some call the manifestation process. Others call it God giving you the desires of your heart.

Question Your Doubts: Ask yourself if there's evidence supporting your fears, and if you are holding yourself to unrealistic standards. Love is divine and will lead you to the eternal, where doubt and fear are not from God.

Act and Face Challenges: Set realistic and achievable goals. Break larger goals into smaller ones that are more manageable steps to build momentum and track progress. Stand for something, or you will fall for anything.

Learn from Mistakes: Analyze failures or setbacks to identify areas for improvement and adjust your approach for future attempts. We are not bad; we are human, and to forgive is divine. I must include a great meme here that is a pic of Jesus wearing a suit with the crown of thorns, holding a large handgun, pointing toward a perceived enemy with a caption that, when you see it below, says it all:

Courtesy: https://www.facebook.com/share/1DuYr9dg9m/

Practice and Persevere: Consistent effort and willingness to keep trying despite challenges are essential for developing new skills and building confidence. Discipline is again the key here as well. A good starting point is a more frequent use of the sacraments like confession, reconciliation, penance, and Holy Communion.

By implementing these strategies and approaches consistently, you can gradually overcome doubts and fears, build self-confidence, and develop resilience to navigate life's challenges. As you more effectively grow in love, charity, respect for nature, and a personal relationship with the divine, it is important to remember the following:

- *Diverse Interpretations:* It is important to remember that the experiences outlined throughout this book are highly subjective and can be interpreted in several ways, ranging from religious and spiritual to psychological and even skeptical perspectives. The key is to find those things that direct you towards love, the eternal, the divine. The divine is what calls us all to rise and meet the challenges of our time head-on.
- *Avoiding Conflation:* While exploring these similarities, it is crucial to avoid prematurely conflating distinct phenomena without sufficient evidence. All this means is that you do not make a mountain out of a mole hill.
- *Keep Things Real:* Seek greater understanding based on love so that your growth guides you to what really matters. Faith and discernment are crucial aspects in the unification of thought in a way that puts love and authenticity first. It is with an authentic love that the way forward will be illuminated for all of humanity concerning how we come out the other side of the coming Earth changes as a people more united, divine, and more focused on the eternal.

Spiritual Disciplines:

A call to the divine and some assistance for answering that call come in the form of spiritual discipline. Many saints and teachings provide such options. The following are just a few for you to consider:

Saint Augustine: As a spiritual guide, St. Augustine tells us that fasting cleanses the soul, raises the mind, and subjects the body to the spirit. The spirit renders the heart contrite and humble that scatters the clouds of sin, and quenches the fire of uncontrolled passions, and kindles the true light of chastity.

Prayers:

"Help me to govern my passions so that my body and soul may give you eternal praise. In Your most holy name we pray, amen."

"I pray that You fill my heart with gratitude for the good in the lives of others and myself. I pray that You teach me to love so that I seek and support the good of others; amen."

"Grant me perseverance in the fight and diligence in habit; amen."

"Help me desire You and the good of my neighbor before any comfort or ease the world has to offer; amen."

It is not required to do all these things. Just do what you can, start slowly, and never stop. Think of minor changes that, when compounded over time, will make a dramatic difference in your spirituality. This will guide you to eternal love, and eternal love will save all of us from the coming end of the age.

Chapter 27
Tying the Bow – Concluding Thoughts

"The Sun, The Cycles, The Science & The Seers" presents a new way to see the world by broadening our frame of reference, proposing that traditional religious encounters and modern ADAs are most likely diverse cultural interpretations of similar trans-dimensional or spiritual experiences. The core argument is for personal growth in love. As we all become more divine, rooted in the idea that moving beyond a rigid, patriarchal interpretation of history and scripture, and embracing a more complete and balanced understanding of the divine, which includes the divine feminine, is a key to rising to the divine eternal that everyone is called to participate in.

Key Insights:

Reframing History and Scripture: We have looked at biblical narratives, like the creation and flood stories, which draw from older Sumerian and Mesopotamian myths, which were later reinterpreted through a developing monotheistic and patriarchal lens. By considering the original, literal meanings of ancient words, a different story emerges that is less about a singular, all-powerful God and more about multiple *"Elohim"* or powerful entities.

The Divine Feminine: A sizable portion of this work revolves around reintroducing the divine feminine principle into a balanced view of God. It suggests that the Holy Spirit, or the Old Testament's personification of Wisdom, known as Sophia or Chokma, embodies these feminine attributes, like nurturing and life-giving qualities.

Challenging Patriarchy: We then attribute the exclusion of the divine feminine and the focus on a solely masculine divine to historical patriarchal distortions and a *"control narrative"* within religious institutions. Recognizing these cultural biases in interpretation can lead to a more accurate and equitable understanding of the divine and gender roles within society writ large.

Unity and Greater Love: The ultimate goal of this work is to demonstrate a unified approach to faith and life that moves beyond a need for control and obedience toward faith based on greater love and understanding. By reconsidering historical and cultural contexts, one can develop a deeper relationship with the creator and a sense of personal purpose.

"The Sun, The Cycles, The Science & The Seers" ultimately argues that a broader, more open-minded perspective, one that incorporates insights from science, ancient history, and different cultural traditions, can help individuals achieve a deeper faith and a more complete, love-centered understanding of the universe and their place within it. It is clear that between church authority, scientific consensus, and the need for more robust proof, as a people, there is not enough time left to get it all together. That is the reason for this book: those who can see more clearly a need to know and are in love, we can become the change needed in the world, despite institutional resistance. The Sun, The Cycles, The Science & The Seers, discusses the connections between ancient myths, religious texts, modern Unidentified Anomalous Phenomena (UAP), and scientific concepts through a lens of reinterpretation that is intended to broaden your frame of reference. This work is advocating for an open mind and a deeper, more complete understanding of creation and one's purpose within it, based on a re-examination of history and the divine feminine. We have explored the idea that ancients, prophets, and modern seers may have described advanced phenomena like flying *"chariots"* or the *"Merkabah"* using the limited language of their time and suggest a potential unity behind these seemingly disparate events. The text focuses on re-evaluating ancient myths, religious texts, and historical encounters like those involving the Anunnaki, Vimanas, and archangels.

You are encouraged to use this new lens of modern understanding of UAPs (Unidentified Anomalous Phenomena) and ADAs (All Domain Anomalies). The book argues for a *"new frame of reference"* that considers these phenomena as potentially connected to both physical reality and spiritual experiences, challenging traditional interpretations, and patriarchal biases in religious narratives, while also introducing their deep connections to the two timelines that await humanity's choice.

The book's central argument is that a deeper, more unified understanding of these shared global histories can lead to a greater faith and sense of purpose for humanity. It encourages readers to be open-minded about how pre-biblical ideas, modern science, and seers' accounts might converge to change our understanding of creation and our place within it. The book, The Sun, The Cycles, The Science & The Seers, proposes that a unified frame of reference can be found by linking modern phenomena like All Domain Anomalies (ADA) and UAP sightings with ancient myths, historical accounts of saints and seers, and scientific concepts like the Sun is affected by cyclical events on a cosmic very long time scale. This synthesis aims to move the reader to a deeper relationship with the creator by integrating these seemingly disparate ideas so that more of us are more likely to choose a path of love over fear, of unity over hate, of heaven over hell, of ascension over devolution.

Core Connections:

Ancient Myths and ADA/UAP: The book draws parallels between historical religious visions and modern UAP/ADA encounters that suggest that accounts of *"flying machines"* or *"chariots"* in Sumerian, Indian Vimanas, Greek, Norse, and Native American cultures were ancient interpretations of advanced, unexplainable ADA phenomena. Biblical accounts, such as Ezekiel's vision and the *"pillar of fire"* that guided the Israelites, are reinterpreted as encounters with a physical, possibly mechanical, flying object known as the Merkabah or *"hornet."*

Science and Spirituality: The book attempts to bridge the gap between science and spirituality by incorporating scientific knowledge of the Sun's magnetic cycles into the broader narrative. It mentions the 12,000-year solar magnetic flip cycle as a known scientific phenomenon. The text implies a connection between these natural cycles, the warnings from seers, and a greater understanding of creation.

Unified Frame of Reference: By considering these varied accounts, from ancient texts and myths to scientific data on solar cycles and modern ADA reports, the author suggests that a *"unified process"* emerges. This unification challenges traditional, often patriarchal, interpretations of history and religion, allowing for a broader perspective that sees all these events as potentially related. The ultimate goal is to foster an *"open mind"* and a deeper, more meaningful relationship with the creator through this expanded understanding, with the hope that humanity reaches a critical mass of awakened souls so that we might enjoy the divine beatific vision, heaven, the higher dimension.

About the Author

Michael is an autodidact who is a student of many disciplines. He received a master's degree in the science of management with a concentration in organizational leadership from Colorado Technical University. He is a practicing Catholic with seven children, and seven grandchildren. He loves family time, fishing, and sharing what he has learned in the hopes that others may find what he has found; a deeper love, greater spirituality, and a larger frame of reference that is going to change the world for the better.

Library of References

1. America Needs Fatima. (2013, June 5). The three days of darkness and prophecies of latter times. https://americaneedsfatima.org/articles/the-three-days-of-darkness-and-prophecies-of-latter-times
2. Alfvén, H. (1942). Existence of electromagnetic-hydrodynamic waves. Nature, 150(3805), 405–406.
3. Alfvén, H. (1950). Cosmical electrodynamics: Fundamental principles. Oxford, England: Clarendon Press.
4. Alfvén, H. (1963). The universe: Plasma and magnetic fields. Reviews of Modern Physics, 35(3), 599–617.
5. Alfvén, H. (1976). Electric currents in cosmic plasmas. Reviews of Geophysics, 14(4), 569–574.
6. Alfvén, H., & Arrhenius, G. (1976). Evolution of the solar system. Washington, DC: NASA Scientific and Technical Information Office.
7. Alfvén, H. (1981). Cosmic plasma (2nd ed.). Dordrecht, Netherlands: D. Reidel Publishing Company.
8. Catholic Answers. (2025, October 29). The three days of darkness.
9. Catholic Online. (n.d.).
10. Cayce, E. (1971). Edgar Cayce on Atlantis. New York, NY: Warner Books.
11. Cayce, E. (1969). Edgar Cayce on Dreams. New York, NY: Warner Books.
12. Carlotto, M. (2022). Toward a New Theory of Earth Crustal Displacement. Journal of Scientific Exploration, 36(1).
13. Cayce, E. (1973). Edgar Cayce on Reincarnation. New York, NY: Warner Books.
14. Cayce, E. (1972). Edgar Cayce on Healing. New York, NY: Warner Books.
15. Cayce, E. (1968). Edgar Cayce on ESP. New York, NY: Warner Books.
16. Cayce, E. (1974). Edgar Cayce on Prophecy. New York, NY: Warner Books.
17. Cayce, E. (1971). Edgar Cayce on Diet and Health. New York, NY: Warner Books.

18. Cronin, R. (2025). Tidal Pumping, Solar Weather, & Geothermal Heat Drive Earth's Weather & Climate.
19. Derek Prince Ministries. (n.d.). Five purposes of Christ's second coming. Retrieved from https://www.derekprince.com
20. Envisioning. (2026, February 15). Project Looking Glass | Xenotech. https://www.envisioning.com/research/xenotech/project-looking-glass
21. Faulkner, R.O. The Ancient Egyptian Pyramid Texts, Oxford 1969
22. GotQuestions.org. (2025, June 23). What does the Bible say about the three days of darkness? https://www.gotquestions.org/three-days-of-darkness.html
23. https://youtu.be/AFwzDpHrOmM?feature=shared
24. https://rumble.com/v414ps7-the-sun-does-nova-in-memory-of-doug-vogt.html?e9s=src_v1_s%2Csrc_v1_s_o
25. https://rumble.com/v52zecf-video-series-part-10-the-source-of-the-electrons-for-the-electric-universe.html?e9s=src_v1_s%2Csrc_v1_s_o
26. https://rumble.com/v416dj8-suspicious0bservers-the-sun-does-nova-doug-vogt-2023.html?e9s=src_v1_s%2Csrc_v1_s_o
27. https://rumble.com/vn0mkv-psychic-focus-on-doug-vogt-shifting-of-earth-theory.html?e9s=src_v1_s%2Csrc_v1_s_o
28. https://www.researchgate.net/publication/342693615_An_analytical_investigation_of_the_theory_of_Douglas_Vogt_regarding_recurring_solar_nova_events_every_12068_years_that_can_explain_the_catastrophic_events_of_the_past_Several_claims_are_difficult_to_p#:~:text=Additionally%2C%20since%20the%20earliest%20sites,Article
29. https://youtu.be/2chUrHZL_ZQ?feature=shared
30. https://youtu.be/le69Jcm-U6E?feature=shared
31. https://youtu.be/4fC25984Xh8?feature=shared
32. https://youtu.be/acqHKS6caw8?feature=shared
33. https://www.youtube.com/watch?v=cEMXp1HlzUs
34. https://youtu.be/cJZkZj4HloY?feature=share
35. https://youtu.be/F-d4zfovcog?feature=shared
36. https://youtu.be/PS_yq19Af-s

37. https://youtu.be/bUsZN4BRE9Y?feature=shared
38. https://youtu.be/MJZsAQHxjcc?feature=shared
39. https://www.catholic.com/magazine/online-edition/will-there-be-three-days-of-darkness
40. https://www.history.com/articles/ancient-sites-spring-equinox
41. https://www.youtube.com/watch?v=IPJy1Z9n6q4
42. https://news.cornell.edu/stories/2024/06/study-confirms-rotation-earths-inner-core-has-slowed
43. https://rotations.berkeley.edu/a-tumbling-t-handle-in-space/
44. https://theethicalskeptic.com/2024/05/12/exothermic-core-mantle-decoupling-dzhanibekov-oscillation-ecdo-hypothesis/
45. Hebrew as the Language of the Proto-Consonantal Script (Carta: Jerusalem, 2016)
46. Hart, George. The Routledge Dictionary of Egyptian Gods and Goddesses, Routledge. 2005.
47. Horanyi, M. (1996). Dust in the solar system and other planetary systems. Reviews of Geophysics, 34(4), 405–431.
48. Izmodenov, V. V., Malama, Y. G., & Zank, G. P. (2000). Interaction of interstellar hydrogen and helium with the solar wind. Space Science Reviews, 91(1-2), 1-134.
49. IOPscience
50. InAirspace. (2026, February 19). Project Looking Glass conspiracy and the battle over our future timeline. https://inairspace.com/blogs/learn-with-inair/project-looking-glass-conspiracy-and-the-battle-over-our-future-timeline
51. Jeff A. Benner; https://www.ancient-hebrew.org/roots-words/root-system-of-hebrew-words.htm
52. Jacobus de Voragine. (1993). The golden legend: Readings on the saints (W. G. Ryan, Trans.). Princeton University Press. (Original work published ca. 1260).
53. Langan, C. M. (2002). Cognitive-Theoretic Model of the Universe: A New Kind of Reality Theory. Self-published.
54. Langan, C. M. (2020). Cognitive-Theoretic Model of the Universe: A Revised Edition. Self-published.

55. Malagnini, L., Dreger, D. S., Parsons, T., Valensise, G., Michelini, A., & De Natale, G. (2025). Do Graviquakes Exist?. Bulletin of the Seismological Society of America.
56. NASA Science. (2009, October 15). The bubble of our solar system. https://science.nasa.gov/resource/the-bubble-of-our-solar-system/
57. Origins of the Hebrews: New Evidence of Israelites in Egypt from Joseph to the Exodus (New Creation: Nashville, 2021).
58. Oxford Research Encyclopedias
59. Private revelation: Prophecy and the Church. https://www.catholic.org/prayers/prophecy.php (General information on the Catholic view of private revelation)
60. Parker, E. N. (1958). Dynamics of the interplanetary gas and magnetic fields. The Astrophysical Journal, 128, 664.
61. Phys.org
62. Scientific American
63. St. Gregory the Great. (2023, June 21). Three days of darkness. https://gregorythegreat.com/three-days-of-darkness/
64. Slavin, J. D., & Frisch, P. C. (2012, November 8). Interstellar dust and the sun. Center for Astrophysics | Harvard & Smithsonian. https://www.cfa.harvard.edu/news/interstellar-dust-and-sun
65. Salter, B. W. (2026). Proposals to Recalibration and Dynamics in Earth's Rotation: Through the lens of ITF.
66. ScienceDirect.com
67. Société des Bollandistes. (1863–1940). Acta sanctorum quotquot toto orbe coluntur (3rd ed., Vols. 1–67). Victor Palmé. (Original work published 1643–1794).
68. Space.com
69. Spreaker. (2025, February 12). Project Looking Glass: A time machine or government hoax? https://www.spreaker.com/episode/project-looking-glass-a-time-machine-or-government-hoax-did-the-government-build-a-time-viewer-using-alien-tech--64335418
70. The Routledge Dictionary of Egyptian Gods and Goddesses, 2005
71. The Global Egyptian Museum

72. Tactical.com
73. The Gospel Coalition. (n.d.). The return of Christ. Retrieved from https://www.thegospelcoalition.org
74. University of Houston
75. Virgo Sacrata. (2023, November 26). Three days of darkness by the prophets and our times. https://www.virgosacrata.com/three-days-of-darkness.html
76. Velikovsky, I. (1950). Worlds in collision. New York, NY: Macmillan.
77. Velikovsky, I. (1952). Ages in chaos. New York, NY: Doubleday.
78. Velikovsky, I. (1955). Earth in upheaval. New York, NY: Doubleday.
79. Velikovsky, I. (1977). Peoples of the Sea. Garden City, NY: Doubleday.
80. Velikovsky, I. (1978). Ramses II and His Time. Garden City, NY: Doubleday.
81. Velikovsky, I. (1977). Stargazers and gravediggers: Memoirs to Worlds in Collision. New York, NY: William Morrow.
82. Vought, D. (2015). The Theory of Multidimensional Reality.
83. Venero, P. L. (2024, December 29). Are we seeing the fallout of 2012's converging timelines? Substack. https://peterluisvenero.substack.com/p/the-mysterious-convergence
84. Vogt, D. B. (1996). Gravitational mystery spots of the United States: Explained using the theory of multidimensional reality. Vector Associates.
85. Vogt, D. B. (2007). God's day of judgment: The real cause of global warming. Vector Associates.
86. Vogt, D. B. (2015). The theory of multidimensional reality (2nd ed.). Vector Associates.
87. Vogt, D. B., & Sultan, G. (1978). Reality revealed: The theory of multidimensional reality. Vector Associates.
88. www.researchgate.net
89. www.youtube.com/watch?v=MrGeReWIHLg
90. Wilkinson, Richard. H. Reading Egyptian Art. Thames and Hudson 1992.

91. Waitz, G. (Ed.). (1878). Liber de apparitione Sancti Michaelis in Monte Gargano. In Monumenta Germaniae Historica: Scriptores rerum Langobardicarum et Italicarum saec. VI–IX (pp. 540–543). Hahn. (Original work published ca. 8th century).
92. YouTube quotes – https://youtu.be/Z88S23RChMs?feature=shared
93. Youvan, D. C. (2024). Dynamic Earth Design: Reconciling Runaway Plate Tectonics and Core-Mantle Decoupling with the Genesis Flood Narrative.

www.ingramcontent.com/pod-product-compliance
Lightning Source LLC
Chambersburg PA
CBHW071439230926
51846CB00001B/288
* 9 7 8 1 9 7 2 9 9 3 4 3 9 *